Cities That Think like Planets

Cities That Think like Planets

Complexity, Resilience, and Innovation in Hybrid Ecosystems

Marina Alberti

UNIVERSITY OF WASHINGTON PRESS

Seattle and London

Cities That Think like Planets was made possible in part by a subvention from the College of Built Enviroments at the University of Washington.

A slightly different version of chapter 6 appears in Marina Alberti, "Eco-Evolutionary Dynamics in an Urbanizing Planet," *Trends in Ecology and Evolution* 30, no. 2 (February 2015): 114–26.

UNIVERSITY OF WASHINGTON PRESS
www.washington.edu/uwpress

LIBRARY OF CONGRESS CATALOGING-IN-PUBLICATION DATA
Names: Alberti, M. (Marina), author.
Title: Cities that think like planets : complexity, resilience, and innovation
    in hybrid ecosystems / Marina Alberti.
Description: Seattle : University of Washington Press, [2016] | Includes biblio-
    graphical references and index.
Identifiers: LCCN 2016001726 | ISBN 9780295996660 (hardcover : alk. paper)
Subjects: LCSH: Urban ecology (Sociology) | Urban ecology (Biology) | Resilience
    (Ecology) | Ecosystem management.
Classification: LCC HT241 .A4353 2016 | DDC 307.76—dc23
LC record available at http://lccn.loc.gov/2016001726

The paper used in this publication is acid-free and meets the minimum requirements of American National Standard for Information Sciences—Permanence
of Paper for Printed Library Materials, ANSI Z39.48–1984. ∞

Contents

Preface

My son's nightlight is a globe. Every evening at bedtime, he makes up a story about the origin of the planet or about its future. All of his stories are imaginary and, at the same time, grounded in his daily experience of living on the planet and in a beautiful city: Seattle. Children have the ability to identify with the world and the objects in it, to acknowledge what supports them, to see the world through many lenses, to tolerate ambiguities, to accept multiple explanations, to experiment with what is possible, and to exist simultaneously in many physical and mental spaces. These qualities are critical to humanity as a whole, if it is to succeed in creating a human habitat of the long now.

Cities now face an important challenge: they must rethink themselves in the context of planetary change. What role do cities play in the evolution of Earth? From a planetary perspective, the emergence and rapid expansion of cities across the globe might be another turning point in the life of our planet. Earth's atmosphere, on which we all depend, emerged from the metabolic processes of vast numbers of single-celled algae and bacteria living in the seas 2.3 billion years ago. These organisms transformed the environment into one where human life could develop. Can humans now change the course of Earth's evolution? Can the way we build cities determine whether we will cross thresholds that might trigger abrupt changes on a planetary scale (Rockström et al. 2009)? Can the rapid development and emergent patterns of urban agglomerations across the globe represent a tipping point in Earth's life, one on the scale of the Great Oxidation (Lenton and Williams 2013)? Will we drive Earth's ecosystems to unintentional collapse? Or will we consciously steer the Earth toward a resilient new era?

The challenge for urban ecology in the next decades is to understand the role humans play in eco-evolutionary dynamics (Post and Palkovacs 2009). Humans are not simply changing ecological conditions globally (Crutzen 2002; Vitousek et al. 1997); we are changing what Hutchinson (1965) called the ecological stage on which the evolutionary play is performed. From a planetary perspective, NASA's Visible Earth Night Lights images suggest an even more extraordinary story of urbanization: the planet and life are co-evolving, changing the courses of each other's histories (Frank 2013). If, as emerging evidence indicates, rapid evolution does affect the functioning and stability of ecosystems (Schoener 2011), current rapid environmental change and its evolutionary effects may have significant implications for ecological and human well-being on a relatively short time scale. Integrating humans into the study of eco-evolutionary feedback can generate significant insights to advance understanding of urban ecosystems' functions and lead to major revisions in the theories of ecology and evolution on a human-dominated planet (Alberti 2015).

A science of cities as coupled human-natural systems has yet to be developed. During the past few decades, we have learned a great deal about how urbanization affects ecological conditions (Grimm et al. 2008a; McDonnell and Pickett 1993; Pickett et al. 2011). Yet the complex mechanisms and feedbacks governing the dynamics of human-natural systems are poorly understood (McPhearson et al. 2016). We do not know how local interactions among human and biophysical processes shape the urbanization patterns of metropolitan regions or how emerging patterns affect human and ecological functions in urbanizing regions. Evidence from growing numbers of studies does indicate that we need to redefine the assumptions of traditional theories and methods in ecology and human sciences if we are to understand such complex dynamics (Alberti 2008; Grimm et al. 2008a; Pickett et al. 2013; Liu et al. 2015). Ecology has long excluded humans from ecosystem studies (Alberti et al. 2003). Thus, ecological experiments conducted primarily in pristine areas can offer designers and planners only limited scientific knowledge, except for the typical advice to keep humans out. By articulating testable hypotheses about the interplay between human agency and eco-evolutionary dynamics, urban ecology has a unique opportunity to advance ecological science and practice (Forman 2014).

In this book, I advance the hypothesis that cities are hybrid ecosystems: the product of co-evolving human and natural systems. The hybrid city simultaneously serves social and ecological functions and is defined by complex interactions among these functions. I ask: What makes an urban ecosystem simultaneously resilient and able to change? Do urban ecosystems have generic properties or qualities that predict their adaptive capacity? What underlying mechanisms explain variation in their ability to self-organize and evolve? How can the science of complex systems help us tackle these questions? Emerging evidence in ecological and social systems indicates that when systems are heterogeneous (i.e., their components vary) and modular (i.e., those components are not entirely connected), they tend to be better able to adapt than those whose elements are homogeneous and highly connected (Scheffer et al. 2012). In hybrid ecosystems, resilience—their ability to adapt to changes—depends on the diversity of biological organisms and on social groups and the economic activities that coexist within them. These ecosystems entail a diversity of cultures and human values as well as the existence of conflicts. Modularity implies loose connectivity among components and network nodes, which allows autonomous functionality. Diversity and modularity support the system's self-organization and provide the flexibility necessary for change. Cross-scale interactions and discontinuities provide opportunities for innovation and point to ways that systems can change and evolve (Holling and Gunderson 2002).

Cities are where innovation has historically occurred. The key role that cities have played in the development of science and technology, and in the generation of inventions and innovations—intellectual and material, cultural and political, institutional and organizational—has been well documented by scholars in a diversity of disciplines (Angel 2012; Glaeser 2012). While rapid urbanization accelerates and expands human impacts on the global ecosystem, it is the close interactions of diverse people that make cities the epicenter of both social transformation and technological innovation (Bettencourt 2013). Yet innovation is tightly linked to the capacity of urbanizing regions to adapt and evolve in a changing environment. For human civilization to achieve its full potential, it is essential to place technological innovation and social transformation in the context of local and global environmental change.

Interdependence between human and ecological processes in cities creates unprecedented challenges for city planners and designers; at the same time, it provides unique opportunities for innovation. This book provides a road map to uncovering the emerging patterns, functions, and dynamics of urbanizing regions. I explore how we can develop an understanding of multiple equilibria and regime shifts in urban ecosystems and how a new planning paradigm can account for these phenomena (Norberg and Cumming 2008). This shift in paradigms will require a new level of integration between the natural and human sciences (Liu et al. 2007a) and between science and design (Pickett et al. 2013) at multiple scales, from the human experience of place (Beatley 2010) to the regional (Forman 2014) and the global (Grimm et al. 2008a). I discuss lessons that urban designers and planners can draw from complexity theory and from the dynamic of coupled human-natural systems, their self-organization, emergent properties, and resilience. Resilient cities require both knowledge and imagination. Planning strategies become reverse experiments to learn how human-natural ecosystems can co-evolve and succeed.

This book proposes a co-evolving paradigm: a view that focuses both on unpredictable dynamics in ecosystems and on social and institutional learning. It develops ecological principles of design and planning: adaptation, flexibility, resilience, and transformation. In the final chapter, I suggest that we need a new ethic: to "build cities that think like planets" so that we might face the challenge of positioning the city in the context of planetary change. For Aldo Leopold (1949), "thinking like a mountain" meant expanding the spatial and temporal scale of land conservation by incorporating a mountain's dynamics. I suggest that we build on Hirsch and Norton's (2012) idea of "thinking like a planet" to expand the time and space dimensions of urban planning to the planetary scale.

Cities are the product of natural and human history and evolution. But they are also the product of human imagination. In Italo Calvino's *Invisible Cities* (1974 [1972]), Marco Polo describes the city of Fedora to Kublai Khan. In his description, the city's museum contains crystal globes that hold miniature representations of the city that individual inhabitants might have developed but never did. Urban ecology, like the map of Kublai Khan's empire, should have room for both the "true" Fedora and the little Fedoras in the glass globes, "not because they are all equally real, but because all are only assumptions. The one contains what is accepted as necessary when it is not yet so; the others, what is imagined as possible and, a moment later, is possible no longer" (ibid., 32). In this book, I propose that we can uncover the fundamental laws that regulate the hybrid city only if we link science to imagination so that we can discriminate between what is probable and what is plausible and learn how to achieve what is desirable for the future of our urbanizing planet.

In this book I ask: What futures are we unable to imagine? I argue that what might be beyond our imagination are cities in which humans are key players in nature's game; cities that bio-cooperate, not simply bio-mimic natural processes; cities that operate on planetary scales of time and space; cities that rely on wise citizens and not just smart technologies. These are what I call cities that think like planets. The emergence of a new urban science that aims to uncover universal rules of how cities work is key to envisioning such transformations. But science and data answer the questions that we are able to formulate. How can we expand our imagination to formulate new questions?

A Road Map

Chapter 1 asks the reader to imagine the future and explores how our imagination of the future can transform the way we live in the present. Building on ecology and evolutionary theory, I contend, in chapter 2, that it is cities' hybrid nature that makes them simultaneously unstable and unpredictable, and also capable of innovating. The chapter examines complexity, emergence, regime shifts, innovation, and resilience in urban ecosystems and the scientific challenges they pose to urban ecology and resilience science. Emerging findings and observations of ecological anomalies in urban ecosystems are difficult to reconcile within current ecological theories; so, as I indicate in chapter 3, they require a new paradigm that better explains patterns observed in urbanizing regions.

Uncertain future interactions between social and ecological dynamics call for a paradigm shift in urban design and planning. Urban ecosystems are qualitatively distinct from other environments. In such systems, change and evolution are governed by complex interactions among ecological and social drivers. In chapter 4, I investigate the emergent properties that characterize urban ecosystems by focusing on patterns (e.g., sprawl), processes (e.g., hydrology), and functions (e.g., flood regulation). I develop an analytic approach by which to examine complex interactions between slow and fast variables that control critical transitions, regime shifts, and resilience. I articulate a set of hypotheses and a research agenda to explore relationships between emergent properties of hybrid ecosystems and their abilities to adapt and innovate.

Studies of complex systems have begun to uncover direct relationships between system structures and resilience. Change, whether gradual or abrupt, is integral to the way nature works. Chapter 5 articulates the hypothesis that variable patterns of urbanization and modular urban infrastructure may be key to cities' resilience. I challenge the assumption that any single optimal pattern of urbanization is consistently more resilient than any other. I use three examples—carbon, nitrogen, and bird diversity—to illustrate the complex relationships between patterns of development and key slow and fast variables that regulate ecological resilience. I suggest that policies and management systems that apply fixed rules for achieving stable conditions by optimizing one function at one scale may make the overall system vulnerable and eventually lead to its collapse.

Chapter 6 specifically focuses on eco-evolutionary feedback. Humans are major drivers of micro-evolutionary change, but, at the same time, completely novel interactions between human and ecological processes may produce opportunities for innovation. Understanding the mechanisms by which cities mediate evolutionary feedback provides insights into how to maintain ecosystem function on an urbanizing planet. To develop and test a theory of urban ecology and the role of cities on a planetary scale, we need to redefine our methods and experiments and rewrite our protocols for collecting and synthesizing data.

In chapter 7, I propose that we define strategies as reverse experiments through which we can learn how urban ecosystems function, co-evolve, and succeed. Yet refining methodologies for studying urban ecosystems does not eliminate the complexity inherent in the fact that our knowledge is inevitably incomplete and that incompleteness, uncertainty, and surprise play a large role in the evolution of scientific thinking and decision-making. We must expand our ability to access multiple and diverse sources of observation and knowledge, as I discuss in chapter 8.

I attempt a synthesis of theory and imagination in chapter 9, where I suggest that by navigating through time, we can uncover our biases about what we know and challenge the idea that there is an end to discovery. I also propose that we can learn from the future: if our cities are to be resilient on a planetary time scale, we must expand our horizons of time and space as well as our ability to embrace change. This chapter discusses the implications of complexity and uncertainty for framing management strategies for the cities of the future, and it articulates a series of principles for urban planning and design.

A Note on Style

The book focuses on the idea of cities as hybrid ecosystems. It is grounded in the science of urban ecology, but it aims to speak to a larger audience of urban designers and planners—and, potentially, to readers outside those fields—about the principles that can transform the way we see and build cities. I see the book as an extended essay: while it seeks primarily to contribute to the science of cities, it develops in several writing styles, ranging from fiction to scientific writing. Chapter 1, for example, describes four fictional images of the future city, while the style of chapter 6 is closer to that of a scientific article. By using these different approaches, I intend to make concepts accessible to a range of readers while maintaining a unified storyline. Each chapter contains explicit cross-references to other chapters, but chapter 10 is the one that brings together all of the book's diverse elements.

Acknowledgments

Many people contributed to this project in crucial ways. My son's curiosity and wonder are my primary sources of inspiration and wisdom. Students, post-docs, and faculty at the Urban Ecology Research Lab, University of Washington, provided me with essential intellectual challenges and insights. Matt Patterson, Tracy Fuentes, Karen Dyson, Emma Solanki, Dorian Bautista, Ashley Bennis, Sarah Titcomb, Yan Jiang, and J. D. Tovey, to name just a few, influenced the most recent evolution of the ideas I present here, but many others, now at other institutions, inspired the explorations and research that led to the creation of this book. They include Daniele Spirandelli, Julia Michalak, Adrienne Greve, Michelle Kondo, Karis Tenneson, Ahmed Al-Noubani, Kuei-Hsien Liao, Vivek Shandas, Jenny Pertanen, Michal Russo, Stefan Coe, Jeff Heppinstal, Lucy Hutyra, Steven Walters, Matt Marsik, and Libby Larson. John Marzluff, Gordon Bradley, Clare Ryan, Craig Zumbrunnen, Eric Shulenberger, and Maresi Nerad were instrumental in initiating, developing, and evolving the urban ecology team with which I have been engaged for more than two decades; I am grateful to them for the exciting and intellectually stimulating collaboration that led me to the research I present in this book. I particularly thank John Marzluff for his motivating conversations and feedback about how cities impact evolution. Thanks also to astrophysicist Adam Frank and physicist Axel Kleidon for expanding my perspective on the roles that humans might play in the universe.

I am also indebted to many other scientists whose scholarship in urban ecology has made this project possible: Stewart Pickett, Mark McDonnell, Nancy Grimm, and Richard Forman provided pioneering research and leadership to the field of urban ecology in the United States, while Herbert Sukopp has been a leader in Europe. Theoretical physicist Geoffrey West, with his pioneering work on the laws that govern complex networks in both natural and social systems, was a key influence on my view of city complexity; his work, along with that of Luis Bettencourt and their team at the Santa Fe Institute, informed and inspired my view of hybrid networks. Buzz Hollings, Stuart Kauffman, Per Back, and Steve Carpenter helped shape my thinking in urban ecology, especially on complex coupled human-natural systems. Thomas Schoener's pioneering work on the interplay of evolutionary and ecological dynamics has inspired me to explore the role that cities might play in eco-evolutionary feedbacks. Kevin Lynch has influenced my ideas about urban design and planning. Virginio Bettini, Larry Susskind, and Paul Ehrlich taught me to challenge my assumptions about how human and natural systems work.

Kim Stanley Robinson's wonderful books inspired me to begin this book with an imaginative scenario—a passage of speculative fiction—and to integrate into it a variety of other writing styles. I am grateful to him for reading and commenting on the chapter. Thanks also to science fiction writer J. M. Sidorova and to science editor Micaiah Evans for their editorial input on chapter 1. I am also grateful to three anonymous reviewers who carefully read an earlier version of the entire manuscript. Their constructive input and encouragement have greatly improved the book.

Several people provided vital contributions to the production process. James Thompson and Michal Russo produced all of the book's graphics and illustrations, helping me to translate complex concepts and data into effective visual representations. Artist Colleen Corradi Brannigan helped me explore the ideas presented in this book through beautiful drawings. My stylistic editors, Micaiah Evans and Helen Snively, edited the book carefully and thoroughly through several readings. Their excellent comments and critical eyes substantially improved the book's writing style and readability. Liz Martell provided additional editing, confirmed the accuracy of references, and formatted the manuscript—all with exceptional dedication and patience. I also thank Regan Huff, my editor at University of Washington Press, for her continuous support and patience; the University of Washington's College of Built Environments for supporting my work; and the Marsha and Jay Glazer Endowed Professorship for providing funds to finalize the book.

I dedicate this book to three important people in my life. My father,

Antonio, and mother, Leda, nurtured my curiosity and instilled, through example, their passion for science and observation of natural phenomena. The most important person in my life is my son, Matteo, who will, together with his generation, imagine, build, and inhabit the cities of Earth's future—cities that think like planets.

Cities That Think like Planets

1

Cities and Imagination

The separation between past, present, and future is only an illusion,
although a convincing one.

—ALBERT EINSTEIN, LETTER TO BESSO FAMILY,
QUOTED IN BERNSTEIN 1989

THIS CHAPTER STARTS BY IMAGINING THE FUTURE. THE PROTAGO-nist stumbles upon four hypothetical futures while riding the New York subway. The last of his stops takes him to Hybrid City, where he engages in dialogue with a city planner. His unexpected detour from the daily commute challenges his assumptions about city planning. The way we think about the future has implications for the questions we ask and the ways we search for solutions. The future we see is constrained by our observations of the past. But the future can look quite different from the past, as is evident from the "great acceleration" of the past half-century. Our imagination of the future can trans-form the way we live in the present.

The pink haze of New York City's twilight, reflected by high-rise build-ings, is the only element of the outer landscape perceptible to Max, who, absorbed in his thoughts, is headed to the subway.

The mayor's budget meeting has left him quite excited, but also unsettled. The mayor announced that over the next three years, he will invest in the creation of a model city. "A city"—the mayor attempted a metaphor—"that thinks like a planet. Next week, I'll hold a town hall meeting with citizens, scientists, and planners." And then he added, before dismissing his budget committee, "I am not looking for answers. We need new questions."

New questions? Max wonders, his mind caught momentarily by the changing shades of blue above the crisp skyline before he descends underground to board his train.

Hybrid Cities

"*Where am I?*" wonders Max as he exits the train at what he'd expected to be his destination. "This is not the right station. Did I get off too early? Too late?"

Each evening, *Use Less*, one of the panels in Jackie Chang's iconic *Signs of Life* mosaics, blazoned on the wall of the mezzanine, welcomes Max to the L platform. But today it's not there. And where he'd expected to see the yellow, beige, and brown Lorimer St. sign, he instead reads, Grand St. Even more surprising is the unfamiliar silence.

The doors slide closed and the train departs before Max has time to consider reboarding, so he starts to walk down the platform. Then he wonders, "Is this a dream?" The question reassures him, frees him to keep walking and thinking. What would happen if one day he exited the subway to find a city different from the one he knew? What would it look like? How would it feel?

And then an ice-cold gust of wind slaps his face. He is outside.

"It's freezing here!" Max exclaims to no one—or to himself. "When did it get so cold?" The streets around him are dark. And quiet.

Glacial City

Max passes a flashing monitor beside the subway exit. *Error*, it pulses, and beside it are the blanks of a dead digital thermometer. The streetlights are flickering; the roads are empty. A single truck occupies the intersection, abandoned. "Where am I?" Max sees a rusted-out newspaper box. He shines his phone's flashlight through the box's shattered window. The front page of a weathered newspaper is dated June 8, 2116. Max picks through the shards of glass to retrieve it. Headlines declare that a new species of aquatic bird has been sighted on the frozen waters near Manhattan. There's a photograph with the story, showing what looks like emperor penguins. "That's impossible," he thinks.

An icy blast rips the newspaper from Max's hands and then blows him down the street after it. The cold is real, brutal. More and more anxious, Max picks up his pace; he almost runs now. He sees the next subway entrance ahead and ducks into it.

Downstairs, a layer of ice, inches thick, covers the floor. The rails aren't even visible—they're completely flooded and frozen over. There are no trains here, and none will arrive. No trains, no lights, no evidence of life—only the artifacts of humanity remain to mark its passing. Max tries but can't stop his teeth from chattering. He stares into the blind hole of the tunnel, and nothing but long, cold darkness stares back. "Please let this be a dream," he whispers.

Empty City

Rumble, clatter. A high-speed train blasts through the tunnel, wrenching Max from sleep. The ice is gone, and the station is lit by scattered ceiling lamps. Max exits the station into the light of an overcast day. Run-down buildings hem an empty intersection. He could not see far last night, but he saw enough to know that the place where he now stands is entirely different. Almost entirely—still no sign of life.

Max begins to walk. A stark outline of empty skyscrapers cuts into the sky. It looks like a movie set for the metropolis of the 1950s: all the action takes place at a bank or a police station, while everything else is a mere backdrop. A feeling of desolation invades Max. His eyes scour the street for evidence of life. If only he could catch a glimpse of New York's familiar face. This emptiness is so unsettling.

Silence gives way to the distant hum of a vehicle. Soon a car comes into sight. It approaches slowly—a gray SUV with reflective windows—and stops at the intersection for a red light. A screeching, grating noise fills the air. Down the street from Max, a gate opens in the side of a building: the entrance to a parking garage. The light changes and the SUV pulls forward, turns into the entrance, and disappears as the gate closes behind it.

Somewhat encouraged, Max begins to walk. He hears noises of heavy construction and goes toward them, then treks through empty streets for a whole mile before finding himself at a familiar site: the East River, at the foot of the Brooklyn Bridge. The tide is high—higher than he has ever seen it. Crews of hard-hatted workers and giant machines are constructing what appears to be an enormous containment wall along the shores, a concrete dike of epic proportions interrupted at key intervals by automated water gates. Just as this city's streets are designed to keep traffic in order, a flood control infrastructure is being built to keep nature at bay.

Max assures himself that he can't really be awake. He watches himself as if from outside, walking across what looks like a movie stage. The Batsignal is projected on the empty metropolitan landscape. Max crosses the bridge and descends into the next subway station to board the first train that arrives. He is its only passenger. The doors close. A recording announces, "Next stop, Columbus Circle." Max exits, hoping to find the New York that he knows. And he knows just where to look: Central Park.

Dream City

As he emerges from the subway, Max reaches to remove his sunglasses, only to discover he's not wearing them. Nothing is shading his view. The unexpected colors and shapes are no illusion or distortion. They are real.

Dark, metallic-gray tree trunks rise all around him, supporting canopies of immense black leaves. An art installation? A holographic projection? No, these are real, and there are so many of them. Artificial trees. But why?

As if in answer to Max's question, someone speaks: "Ten times as efficient as the originals, you know."

He turns to find a man with cropped white hair leaning against a nearby newspaper stand.

"I'm sorry . . . what?" Max asks, startled.

"You're a visitor, yeah?" says the man.

"I don't know what you're talking about," says Max. "I've lived here all my life."

"Not to worry," says the man, as though he hadn't heard Max. "If you don't like what you see, you can always move on."

"Where am I?" Max asks.

"Central Park, of course. New York City. Big resilient Apple."

"So these trees are . . . silicon?" Max hazards.

"You guessed it," says the man. "Genetically engineered. Those beautiful black leaves up there—they convert the sun's rays to chemical energy. Just like green plants. But better."

"They look so real," Max exclaims. "Just like trees, except for the color."

"Real? Of course they're real. And they've let us free up nine-tenths of the land we once devoted to biomass production."

"So . . . you live around here?" Max asks.

"Yeah, I moved here five years ago. This is a special place. It's a true carbon-neutral city. Imagine that: carbon-neutral. Smart tech optimizing energy and water usage everywhere you look. Waste—heck, inefficiency of any sort—is almost unheard of. Goes for traffic, too. We've got sensors installed across the city capturing real-time data and feeding it back to traffic signals to regulate vehicle corridors. Even social life is updated in real time, and you can create your own imaginary community or join one ready-made. Even pipe it to your lenses and overlay it on the real—turn this city into whatever you want it to be, just for you."

"Wow, that seems terrific." Max wants to learn more but has trouble even imagining what to ask.

"Yes, it's all good," the man responds, "but this is an unfinished project, an incomplete painting. Always a work in progress. And honestly, living in a smart city can be hard. Everybody trusts technology, but trusting people . . . now, that's something else."

Hybrid City

A feeling of movement wakes Max. He opens his eyes to find that he's no longer outside but back in the train, with glimpses of concrete tunnel wall zipping past the windows. Abstract tones piped through hidden speakers weave themselves into an ambient musical backdrop. Floating in the air before him is a semitransparent, three-dimensional image of Planet Earth. As Max watches it turning on its axis, it slows, then swells, showing North America, the Eastern Seaboard, New York State, Manhattan, and, finally, a flashing blue dot moving through a map of the city. The dot drags along a *You Are Here* banner. And then the map disappears and all that remains is the location indicator. *You are here.* Another word appears: *Now.*

You are here now.

"But *where?*" Max says aloud. "And when?" He begins to feel as if he is simultaneously in multiple spaces and times. But there is no scale of time or space to calibrate his visions. He is at once in Manhattan right now and on Planet Earth a hundred years ago and a hundred years in the future. He is in Manhattan and in Cape Town and in a Vietnamese village and in ancient Thebes. Perhaps an external observer could distinguish distance or direction in the overlapping images that appear in Max's vision, but Max is comfortable exploring such space without any need for direction.

The train decelerates, enters a station, stops. The doors open. The words *Times Square* flash in the air of the station. Max's lightheadedness and confusion give way to wonder as he climbs the stairs to the street.

Outside, an impossible landscape has replaced his city. It is both familiar and unknown. Times Square is a red-maple swamp. A loud *thwack* draws his attention, and Max turns in time to see a beaver swim off toward the opposite bank of a wide, leaf-strewn pond. As Max watches the beaver move through the water, ripples spreading out behind it, he feels a sense of freedom. Noisy birds and beautiful plants take front stage in this realm, as it slowly and seamlessly dissolves into a busy, vibrant, populated place. People invade the streets, flood the square, and swarm in and out of taxis, subway stations, and buildings in apparent chaos. But they never collide and never seem to pause or yield right of way: their movements harmonize perfectly, effortlessly. The whole place is noisy, smelly, colorful, and fast-moving, and it grows noisier and faster still—a crescendo of human activity that seems ready to explode. But as the show nears its climax, the image transforms into a calmer scene, where conversations mesh with birds' songs in a mellifluous soundscape. And choreographed to this music, as in an ancient ritual or a modern dance, people, buildings, and trees improvise a hybrid play of work and motion.

"You are in a hybrid city." The lightly accented voice comes from a young girl with long black hair who has sneaked up beside Max and is now admiring the view. Then she turns to meet Max's eyes and, holding out a tray of packets, says, "Have a maple sugar crisp. They're printed locally, in my neighborhood. My mother wrote the recipe, and we sell the code worldwide. They're printed everywhere now, but mine are best— proprietary tweaks from the author that you won't taste anywhere else." Max reaches for the crisp and nods his thanks.

As he looks into the distance, he notes a sign indicating the way to City Hall—a point of familiarity, perhaps, in this mysterious place. Maybe there he will find an explanation.

City Hall

Max is staring at a beautiful glass building. *City Hall* is painted across the arched entrance in shimmering, changing letters that sample and incorporate the faces and garments of the passing crowd. People move in and out through multiple revolving doors; they sit at tables in the square around the building; they mill about in the foyer, all engaged in conversations, communicating by voice, gesture, and device linkage. As Max approaches the entrance, a message scrolls across his vision: *Welcome to Hybrid City Hall!*

Max heads inside with the excitement and curiosity of a child set loose in a toy store, not sure where to head first. The internal walls are glass, too, so he can see deep into the building, see people working at terminals, meeting with officials, holding debates, giving and listening to lectures. As Max surveys the space, more words appear in the air to inform him about what he is seeing.

Fedora, Max reads on the far side of the foyer. Fedora! One of the Invisible Cities that Marco Polo describes to Kublai Khan in Max's favorite book by Italo Calvino. As Max enters the Fedora room, an array of images appears. Each one expands and contracts and seems to play out in fast forward or slow-motion. As in the fictional Fedora's city museum, Max realizes, these images are representations of the city that its individual inhabitants have imagined and might have developed—the could-have-been cities that never were. A line of text scrolls continuously around the walls of the room: "There is nothing in the form of the laws of nature at the fundamental microscopic level that distinguishes a direction of time. They are time-symmetric" (Barbour 2014: n.p.).

Planning

In the atrium, Max happens upon an interview in progress. "Everything is possible," says the city planner to a reporter from a local TV station who is asking about how the city should respond to the uncertainties of climate variability. "Uncertainty is what makes planning so challenging," the planner continues. "And so exciting. Hermann Weyl [1949: 116], a mid-twentieth-century physicist, said, 'The objective world simply is, it does not happen.' Although the universe is laid out in time and space, past and future are as real as the present."

"Interesting," says Max. "How do you plan in such a world?"

"You plan simultaneously at all scales of space and time," the planner replies,

"Because they all might possibly occur?" says the reporter.

"No. On the contrary," says the city planner, "none of them will actually occur. All of them are based on limited and necessarily incomplete sets of assumptions. The real future that we experience will be something different, but as we prepare our city for every collectively imagined scenario, we shape ourselves into a resilient city able to withstand whatever our ultimate reality delivers."

Suddenly Max realizes that where he is now—Hybrid City—is the collectively imagined city. As in Calvino's *Invisible Cities*, Max recalls, as he wanders off into the flowing humanity outside Hybrid City Hall, the real Fedora is "what is accepted as necessary when it is not yet so"; while the plausible futures in the crystal globes of Fedora's Museum represent "what is imagined as possible and, a moment later, is possible no longer" (Calvino 1974, 32). "On the map of your empire, O Great Khan," Marco Polo tells Kublai Kahn, "there must be room both for the big, stone Fedora and the little Fedoras in glass globes" (ibid.).

Max wakes as he tumbles gently from his plastic seat in the empty subway car. Doors open to a platform and an amplified voice intones, "End of line. Please exit." Max's eyes are invaded by the light of the subway station. The good-humored gaze of an old woman meets his surprised stare.

She smiles and says, "Do you have a question?"

2

The Hybrid Ecosystem

A Hypothesis

URBAN ECOLOGY FACES A SIGNIFICANT CHALLENGE: TO POSITION *itself in the context of planetary change and to understand the role that cities play in the evolution of Earth. The dynamics of urban ecosystems are complex and hard to predict because we do not yet understand how they work. Here I advance the hypothesis that cities are hybrid ecosystems, the products of co-evolving human and natural systems. Building on ecology and evolutionary theory, I contend that it is cities' hybrid nature that makes them unstable and unpredictable, and simultaneously capable of innovation. This chapter examines complexity, emergence, resilience, and innovation in urban ecosystems and the scientific challenges that they pose to urban ecology.*

A science of cities as novel ecosystems has yet to be developed. Cities are the most visible signature of the Anthropocene: a new era in the co-evolution of life and the planet (Crutzen and Stoermer 2000; Monastersky 2015). Yet our understanding of how cities, as coupled human-natural systems, emerge, grow, and evolve is at best tentative and fragmentary. Despite remarkable progress in the study of urban ecosystems over the past few decades, urban ecology still lacks a theory of human habitat that is comparable to ecological theories of natural habitat. What makes the city an optimal habitat for people and their communities? Do thriving cities share certain properties? Can we identify a set of features that explains these cities' vitality and well-being? Are there factors, such as size, density, or form, that best correlate with the health of cities? What makes cities able to adapt, change, and evolve through time? What makes a city resilient?

The study of cities has evolved through many theories, myths, and paradoxes. Cities have been compared to living systems (Geddes 1915), biological organisms (Odum 1971), and ecosystems (Odum 1975). Kevin Lynch (1981) was among the first to point out that theories of urban genesis reflect dominant images of the evolving conception of the relationship between humans and nature. Cities mimic biological systems in many ways, yet they exhibit characteristics that break many known natural rules. More than half a century ago, Jane Jacobs (1961) referred to the city as "organized complexity." Since then, the view that cities are complex systems has emerged as a new challenge for science (Batty 2005). Like living systems, cities are diverse and complex and yet governed by extraordinarily simple universal laws (Bettencourt et al. 2007).

More recently, Bettencourt (2013) compared cities to stars: as the spatial and temporal scale of cities increases, social interactions within them intensify. This process is analogous to nuclear reactions occurring within stars: the larger the population of a city, the higher the rate of its social interactions. Like large stars that burn faster and brighter than smaller ones, cities with larger populations attract even more people, and their social interactions accelerate (ibid.). Yet scientists recently have found that the structure of human interactions in cities has some unexpected underlying qualities. On average, as the size of a city doubles, the total number of social interactions within it *more* than doubles—and in a predictable way—but, regardless of a city's size, the group clustering of its social networks does not change. Whether they live in large cities or small villages, people form tight social communities (Schläpfer et al. 2014). The main difference is that in small towns, interpersonal connections might be defined by spatial proximity, while in large cities, people form social communities through a diversity of communication networks based on affinities and/or interests.

Scientists studying cities as complex systems are curious about the universal principles that govern them and which could lead to a predictive theory of urban growth and change to inform decisions about their management. A team led by Luis Bettencourt and Geoffrey West at the Santa Fe Institute (SFI) has been working toward formalizing a new quantitative understanding of urban function (Bettencourt and West 2010). By examining a large set of data on diverse aspects of urban regions, they observed that cities exhibit scaling relationships similar to those that biologists have found for organisms' molecular, physiological, ecological, and life-history attributes (Bettencourt et al. 2007). And emerging evidence suggests that such phenomena might not be limited to modern settlements. By examining archaeological settlement data from the pre-Hispanic Basin of Mexico, Ortman et al. (2015) found that scaling relationships between socioeconomic production and settlement size have characterized human settlements throughout history.

In biology, the best-known allometric (scaling) relationship is that of an organism's size to its metabolic rate (West et al. 1997); such a relationship influences organisms' rates of growth, reproduction, and mortality. Biologist Max Kleiber (1932) observed that an animal's metabolic rate scales to the three-quarters power of its mass. Large animals consume more energy than smaller ones, but they are also more energy-efficient. This scaling relationship is manifest in the structures and functions of living systems at all levels of biological organization, from populations to ecosystems, as they, too, are influenced by organisms' metabolism (Anderson-Teixeira 2009). Furthermore, researchers have observed a power-law functional form (i.e., an exponential relationship) that relates the area occupied by a biome to the number of species that it supports; because this relationship transcends specific details of organism interactions, it provides important insights into the universal laws that govern the emergent patterns of biodiversity (Martin and Goldenfeld 2006).

Power laws have been observed in both natural and social phenomena. They indicate that these phenomena are scale-free—not associated with any particular scale—so they possess the same statistical properties across all scales. Applied to cities, this concept implies that the ratio of any two cities' input (e.g., energy) and output (e.g., income) is a function of the ratios of the sizes of their populations, but is also independent of the specific sizes of their populations. The SFI team has substantiated the hypothesis that, like biological organisms, cities scale up. As they grow, they become wealthier, more efficient, and more innovative. As in the metabolism of an organism, cities exhibit economies of scale: the bigger a city is, the less infrastructure it has per capita. However, as Bettencourt et al. (2007) pointed out, some socioeconomic quantities have no analogue in biology. Many properties of cities—from patent production and personal income to electrical cable length—are shown to be power-law functions of population size, but the exponents involved in scaling fall into distinct classes. When researchers measure wealth creation and innovation, they find increasing returns, but when they measure infrastructure, they find economies of scale (figure 2.1). While an organism's metabolic rate is inversely related to its size, the opposite is true of cities: bigger cities run faster than smaller ones (ibid.). The pace of social life, as measured by density of social interactions, wealth generation, and speed of walking—but also by incidence of crimes and diseases—is faster in large cities than in small ones.

There is now ample evidence that important properties of cities across diverse world regions increase, on average, faster (socioeconomic superlinearity) or slower (material infrastructure sublinearity) than city population size. More recent studies have suggested that factors associated with urban heterogeneity might explain the variability observed across cities of a given size (Pan et al. 2013; Sim et al. 2015). These effects might also be explained by urban scaling theory in terms of the emergent properties of socioecological networks and built spaces of urban ecosystems.

Analogies have proven powerful in advancing scientific understanding by providing effective heuristics and creative strategies for solving scientific enigmas. They are also especially powerful in drawing our attention to their limits in explaining specific phenomena. Scaling and the discovery of power laws in nature have exposed the underlying dynamics and structures of a large spectrum of scale-free phenomena. Scaling relationships are a useful tool for understanding the structure and dynamics of complex ecological systems and offer clues about the underlying mechanisms governing system dynamics (Milne 1998). Scaling laws tell us that many unique human-natural systems produce universal behaviors. But that is only part of the story of how cities work.

Despite these commonalities, today's urban regions are highly diverse in physical structure, social organization, information flows, biophysical environments, and geographical and political contexts. What I find most intriguing are the phenomena with scaling components that vary between biological and human systems, those that vary among properties within cities, and those that do not fit into power laws at all. Why do cities exhibit "superlinear" relationships while organisms have "sublinear" ones? The super- or sublinear nature of scaling in cities has profound practical consequences for growth and sustainability (Bettencourt et al.

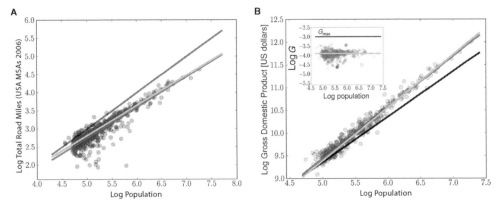

FIGURE 2.1 City scaling. Quantities reflecting wealth creation and innovation have β ≈1.2 >1 (increasing returns), whereas those accounting for infrastructure display β ≈0.8 <1 (economies of scale) (Bettencourt et al. 2007; Bettencourt 2013). Source: Bettencourt 2013, 1438.

2007; West, Brown, and Enquist 2001). Or consider two cities that scale up similarly for certain qualities (e.g., income) but diverge for others (e.g., levels of pollution). How can we explain such divergences? And which qualities of cities cannot even be captured by scaling laws?

To answer these questions, we must first understand the underlying mechanisms of the observed scaling relationships. Power laws are mathematical descriptions of natural patterns, but they do not provide a scientific explanation of the underlying mechanisms. Brown et al. (2002) pointed out that power laws in ecosystems can be used inductively, as empirical patterns, to formulate and test hypotheses about the mechanisms governing ecosystem functions, such as biodiversity. Similarly, scaling in cities, and divergences from scaling laws, can provide useful clues for investigating how, in urban ecosystems, universal principles emerge from biological, physical, and social laws.

Observed patterns in nature and social systems emerge from the complex interactions of diverse components through multiple networks that operate according to a set of basic principles. The composition of biological communities is the result of local interactions between species and their environment via multiple networks (e.g., pollination and food webs). Cooperation in human communities emerges as individuals and organizations interact in a variety of social networks. Both natural and social systems are organized as networks (Barabási and Oltvai 2004;

Ulanowicz 1986). Networks consist of nodes or vertices and the relations (e.g., links or edges) among those components. In biological systems, networks can represent multiple levels of organizations, from genetic interactions to interacting species in communities. In human systems, networks may represent individuals, communities, and organizations interacting through diverse activities (e.g., production, transportation, and communication). The structures and dynamics of networks emerge from interactions between their various elements and determine the rates at which energy, materials, and information are processed.

Hybrid Networks

In studying biological systems, the physicist Geoffrey West (1997) proposed that life is sustained by hierarchical branching networks that are optimized by the process of natural selection. West suggested that quarter-power scaling—the power exponent of three-quarters that is predominant in biology—reflects the fractal-like designs of resource distribution networks (West, Brown, and Enquist 1997, 1999). All biological systems, at various scales of biological organization, depend on the close integration of numerous subunits to function; they are supported by a complex network that supplies resources, processes waste, and regulates systems' functioning. West and Brown (2005, 1578) proposed that natural selection solves this problem by evolving hierarchical fractal-like branching networks that distribute energy and materials between macroscopic reservoirs and microscopic sites. The network performs best when it uses the least energy and materials to distribute resources (West, Brown, and Enquist 1997).

As is true for biological systems, human communities depend on complex networks to distribute energy, materials, and information. Evolutionary anthropologists are curious about whether human social systems self-organize in response to optimization principles similar to those found in natural systems (Burnside et al. 2012). Markus Hamilton et al. (2007) found that the structural properties of hunter-gatherer societies are remarkably similar across a wide variety of cultures and continents. The symmetries that researchers have observed across cultures, environmental conditions, and histories suggest the presence of universal laws governing the ways that human societies self-organize through social networks to maximize the flux of energy, materials, and information (Burnside et al. 2012). Emerging evidence also shows that social networks facilitate the emergence of cooperation, a fundamental aspect of biological networks (Apicella et al. 2012; Ohtsuki et al. 2006).

A key question is whether human societies exhibit unique properties. Sociologists and anthropologists have long been interested in the role that social networks play in the functioning and evolution of human communities. More recently, economic scholars have explored how social networks evolve in relation to economic development and how they affect the creation of social capital and rates of economic growth (Goyal 2015). As societies transition from hunter-gatherer to agrarian to industrial, new social networks have emerged, significantly modifying both the patterns of and the opportunities for social interactions, with significant effects on wealth production and human well-being. But we do not yet know which mechanisms emerging networks use to regulate the functioning of social systems in relation to natural systems. How do the social and economic transitions associated with urbanization increase use of resources, and how does technological development affect, and how is it affected by, human-natural interactions?

Urban ecology's fundamental challenge is to understand the nature of different networks in coupled human-natural systems and to uncover the principles that govern their evolution (figure 2.2). A key question is whether social networks are qualitatively different from ecological ones. Do people exchange information differently than other elements in living systems? And if so, how does human distribution of resources and information differ from that of other organisms? How do different networks cross scales? Furthermore, can we characterize the structure and dynamics of hybrid networks emerging from human-natural interactions within and across temporal and spatial scales? Finally, how do these diverse networks and emergent properties enable hybrid systems to navigate change and evolve?

Such questions are both exciting and daunting. Imagine the diverse range of scientists who would need to collaborate in this exploration, and the new knowledge that could emerge as we try to answer such questions. First, we would need to resolve some enduring scientific debates. Natural selection, for example, is a well-supported mechanism, and evolutionary biologists have long posited that it characterizes the evolution

Ecological Networks	Human Networks	Built Networks	Hybrid Networks
Physical networks (e.g., rivers)	Social networks (e.g., tweets)	Transportation (e.g., roads)	Socio-ecological networks (e.g., conservation networks)
Mutualistic networks (e.g., predator-prey networks)	Economic networks (e.g., financial)	Power (e.g., power grid)	Built-natural networks (e.g., green infrastructure)
Food webs (e.g., salmon)	Information networks (e.g., the Internet)	Water (e.g., water pipes)	Novel food webs (e.g., synantropic species)
Genetic networks (e.g., microbial genetic networks)	Institutional networks (e.g., healthcare, emergency service)	Technology (e.g., fiber)	Emerging eco-evolutionary networks (e.g., novel seed dispersal pathways)

FIGURE 2.2   Emergent networks. Hybrid networks in urban ecosystems emerge from interactions among human networks, ecological networks, and built networks. Photographs: river: davebloggs007; New York City tweets: Eric Fischer; roads: Pratik Gupte; conservation networks: U.S. Fish and Wildlife Service, Pacific; eagle: Jason Mrachina; freighter: steeedm; power grid: Michael Foley; green infrastructure: C. Stephen Haase/TNC; Kokonee salmon: Tim Rains (U.S. Forest Service); Internet: Michael Coghlan; water pipes: Bob Duran; crow: Nottsexminer; photomicrograph of cyanobacteria, Matthewjparke; helicopter: Dave Herholz; fiber-optic cables: Roshan Nikam; seeds: Aelwyn.

of biological networks. They disagree on the role that self-organization (i.e., spontaneous emergence of order in time and space) might play, though. To understand how evolution constructs the mechanisms of life, molecular biologists would argue that we also need to understand how genes affect the evolution of cellular processes, helping to develop elaborate self-organizing rules that constrain the scope of evolutionary change (Johnson and Kwan Lam 2010).

The debate surrounding the architecture and evolution of biological networks has important implications for understanding social-ecological networks. Comparing and contrasting the architecture of and dynamics between social and ecological networks are especially important, as are identifying and describing potential interactions among these elements as they occur in cities. Though few empirical studies have focused on similarities and differences across social and ecological networks, scholars have begun to advance diverse hypotheses about what differentiates them from one another (Janssen, Bodin, and Anderies 2006). West et al. (1999) have shown that in biological networks, properties such as flows of energy and resources systematically increase in a self-similar fashion through the networks, leading to economies of scale (sublinear scaling). In contrast, in social networks, variables such as the strength of social interactions and flows of information are disproportionately higher among individuals than among large clusters, resulting in the observed trend known as superlinear scaling (Bettencourt et al. 2007).

This diversity and complexity in interactions among human communities across the world pose great challenges as we attempt to uncover the properties of social networks that support human resilience and well-being. Social scientists have yet to develop a theory that explains the emerging contradictions between rapid increases in social interactions and the observed variations in different social communities' abilities to thrive. Ernstson et al. (2010) argued that, unlike ecological networks governed by natural selection, people construct the channels through which energy and matter flow in a recursive communication process of alignment and coordination. This could explain, for example, why it is so difficult to disentangle relationships between the ways that social interactions evolve in informal settlements in large urban agglomerations and their capacity for innovation. Under which conditions do increasing social interactions drive social experimentation rather than create poverty traps?

One fundamental question is how human societies evolve and change in response to constraints and feedbacks from their environment. How are new resources and capacities created, and how do novel structures and behaviors arise? How do resilience and transformation emerge in human social systems as compared to natural systems? Young et al. (2006) suggested that in biophysical systems, new structures emerge as system elements interact autonomously, creating novel configurations and shifting between multiple stable states. Similarly, in social systems, structural transformations are emergent properties of agent interactions, but strong elements of purposive "shaping" also occur through planning (ibid., 311). In planning their behavior, social actors operating within institutions act in anticipation of change and in response to social learning. Young et al. pointed out that although other social animals also exhibit foresight and reflexivity, these attributes are central to the adaptive capacity, resilience, and innovation of human social systems.

These observations provide an interesting avenue for exploring the interactions and emerging structures of coupled social-ecological networks. How do the emergence and rapid development of cities across the world affect their evolution? As noted earlier, scientists exploring scaling relations in social networks have shown that the number of social contacts grows with a city's population, which explains superlinear increases in innovation and economic outputs; meanwhile, the average communication intensity among individuals, and the probability that an individual's contacts are also connected with one another, remain constant, indicating that urban social networks retain much of their local structures as cities grow (Schläpfer et al. 2014). Furthermore, the nature of social interactions (e.g., the predominance of long-distance ties in large agglomerations) or characteristics of the population (e.g., the proportion of more highly educated individuals) could provide an explanation for higher levels of innovation in large as compared to small agglomerations (Arbesman, Kleinberg, and Strogatz 2009).

There is no doubt that social interactions play a key role in the underlying dynamics of city growth and evolution. Cities are, most of all, networks of people. It is through social interactions that people have reconfigured ecosystem patterns and processes to create the human habitat. Social networks are human constructs that reflect values, worldviews, ideas, and knowledge, and that shape the institutions that govern them (Van der Leeuw 2007; Ernstson et al. 2010). But we do not know how values, ideas, and knowledge shape humans' interactions with ecological networks and govern their dynamics.

Changizi, McDannald, and Widders (2002) found that differentiation in complex networks increases with network size and that this relationship is consistent with a power law. Their results are explained by the hypothesis that because nodes are costly to build and maintain in such complex networks, the networks optimize their size in some way. But networks that are governed by natural selection (e.g., organisms, ant colonies, and nervous systems) are far better able to combine nodes of different types than are networks that require human ingenuity (e.g., electronic circuits or businesses). This means that networks governed by natural selection use nodes in a combinatorial way to construct a specified function or expression (from genes to ecosystems), but creating a new node requires that more components interact in a coordinated way in order to implement that function. In comparison, humans can quickly build a new electronic component type or add a new employee to an organization: humans are plastic and inventive and can easily increase the number of node types in systems based on their own ingenuity. But humans build networks (e.g., businesses or electronic circuits) that are relatively low in terms of combinatorial expressions—there is only a single node for each given expression (Changizi, McDannald, and Widders 2002).

I propose that in coupled human-natural systems, networks are not governed by either natural selection or human ingenuity alone. Emerging networks are hybrid and novel expressions, and their functions emerge from interactions among natural and social networks (see figure 2.2, earlier). Several types of new hybrid networks may emerge: (1) coupled social-ecological networks linking social process to protected areas; (2) artificial networks that people create to support human activities; (3) novel networks created by interactions among human and natural flows of information, energy, and materials; and (4) eco-evolutionary networks. The multiple functionality of green infrastructure is an example of the complex and diversified functions of networks in hybrid ecosystems. Green corridors in cities may be designed to provide habitat for wildlife and simultaneously mitigate stormwater flows, reduce heat, cool the atmosphere, and absorb atmospheric pollutants. Green infrastructure supports human health and also provides opportunities for recreation (Tzoulas et al. 2007) and ecosystem stewardship (Andersson et al. 2014). The creation and maintenance of multifunctional green space in urban ecosystems require the cooperation of multiple agents, all of which must also synchronize with the natural processes that govern them.

Cross-Scale Interactions

Cross-scale interactions between social and natural systems pose additional challenges to understanding mechanisms governing the dynamics of urban ecosystems. Cities affect and are affected by global economic and ecological changes through social and ecological networks that operate across scales. Interactions across scales in ecological and social networks have been studied, but the study of "tele-coupled" human-natural systems (i.e., those that incorporate distant socioecological interactions) is only now emerging (Liu et al. 2013). While most of our attention is focused on potential consequences of the accelerated pace of change, cross-scale interactions may provide unexplored opportunities for innovation (Westley et al. 2011).

No simple relationship exists between the structure of hybrid networks in urban ecosystems (e.g., their diversity of components and degree of connectivity) and their dynamics or robustness. How do emerging interactions among diverse socioeconomic and ecological variables affect the ability of urban ecosystems to navigate change and evolve? Do complex hybrid networks share emerging principles that govern resilience in other complex ecological and socioeconomic networks? Which characteristics and feedbacks emerge from human-natural interactions in such settings that might differentiate hybrid ecosystems? And to what extent does a special set of interactions across time and space generate a new level of heterogeneity? Each city is a unique phenomenon emerging from interactions between human and ecological communities, as shaped by their biophysical setting and their cultures and histories.

Scientists have made important progress in their efforts to uncover relationships between complex network structures and their resilience (Scheffer et al. 2012), providing fertile ground for further exploration. Robustness in hybrid networks is typically associated with heterogeneity and modularity, as exhibited by resilient socioeconomic and ecological systems. These networks are disassortative (nonrandom): highly connected large nodes are disproportionately connected to small nodes. By examining pollinators, for example, Bascompte, Jordano, and Olesen (2006) showed that asymmetries inherent in co-evolutionary networks may confer a significant degree of stability against disturbance and facilitate biodiversity maintenance. Variability in the degree of interconnectedness among elements allows semiautonomous functionality and robustness. Yet robustness also depends on the quality of the highly connected nodes. Drawing lessons from financial markets, May, Levin, and Sugihara (2008) demonstrated that some types of compartmentalization can be harmful, in that they may preclude stabilizing feedbacks.

We can expect that hybrid networks' responses vary from those of other networks because heterogeneity and modularity involve trade-offs between local and systemic risks. Studies of biological, technological, and other types of networks show differential structural and dynamic robustness (Solé and Montoya 2001). While quite robust against random errors, highly heterogeneous and modular networks are extremely fragile to attacks targeted at hubs. These insights are important when examining the behavior of dynamic networks. Financial and other heterogeneous networks, such as the Internet and other social networks, are especially vulnerable to attacks that target their highly connected nodes.

Another important distinction between hybrid human-natural systems and socioeconomic or ecological systems results from interactions among different network topologies. By examining financial systems, for example, scientists have shown that payment systems may not always be the relevant network for understanding systemic events in liquidity transactions (May, Levin, and Sugihara 2008). Political and social networks may play a key role and could ultimately influence panic behaviors in response to a financial crisis. As noted by Levin, who compared the management of the electric grid to that of financial systems, robustness in financial systems is an emergent property, "It cannot be engineered" (quoted in Kambhu, Weidman, and Krishnan 2007: 48). However, policymakers use top-down interventions to modify individual behaviors in an effort to support the collective good. But in coupled human-natural systems, it is not practically possible to control individual perceptions and behaviors in a top-down manner.

Extending the study of ecosystems to human-dominated systems has been crucial for advancing our understanding of urban ecosystems. During the past few decades, we have learned a great deal about which factors control the development of urbanizing regions and their environmental conditions, but we are only now starting to understand which elements influence their resilience. As we have learned more about urban ecosystems, we have inevitably become less confident about what we know and can accurately predict about the future of such systems. Empirical observations have revealed paradoxes and contradicted earlier generalizations, uncovering the limitations of our scientific assumptions and approaches. What makes cities unique ecosystems is human agency.

Resilience in urban ecosystems is governed by the interactions among multiple agents, and among them and the patterns, processes, and functions that operate across multiple scales. Adding human-ecological interactions to ecosystems introduces many more dimensions of complexity to system dynamics and resilience. One such dimension is foresight, a human capacity that is not observed—at least not at the same level of complexity—in nonhuman species. Understanding how cities emerge, grow, and evolve requires us to both expand our scientific knowledge and broaden our imagination.

A Science of Cities as Hybrid Ecosystems

I advance the hypothesis that cities are hybrid ecosystems: the product of co-evolving human and natural systems. From an ecological viewpoint, they differ markedly from historical ecological systems: their species composition and relative abundance have not occurred before within the same biome (Milton 2003). But urban ecosystems also differ significantly from historical human settlements: they are novel habitats and contain both natural and human historical features. They are hybrid systems—simultaneously unstable and capable of innovating.

In ecology, the concept of "novel ecosystem" has evolved since the beginning of the past century. Mascaro et al. (2013) suggested that novel ecosystems occupy spaces that are far removed from historical ecosystems, in abiotic (e.g., climate) and/or biotic (e.g., species-competition) terms. A foundational concept is Gleason's (1926) idea that communities emerge as a result of species' individual responses to a continuously changing environment. Complementing Gleason's idea is our increasing understanding of the biotic and abiotic interactions that govern ecosystem dynamics. A greater emphasis on the role of human agency as a source of "novelty" was provided by Chapin and Starfield (1997), who defined a novel ecosystem as the result of anthropogenic changes.

In an attempt to synthesize the diverse perspectives, Hobbs et al. (2006, 2) defined two key elements that characterize novel ecosystems: "(1) novelty: new species combinations with the potential to change ecosystem functioning; and (2) human agency: ecosystems that are the result of deliberate or inadvertent human action, but do not depend on continued human intervention for their maintenance." Novel ecosystems are not recent phenomena or necessarily the result of human intervention. What *is* new is the acceleration of processes that give rise to novelty and the dominant role of human action: both have significant implications for ecosystem management. Earth scientists are providing increasingly more evidence that anthropogenic changes are key drivers of novel ecosystems. Yet the extraordinary implications of this shift in the history of our planet's evolution have not been explored. Until recently, ecologists have characterized human agency as an external driver of change through activities including land conversion, resource extraction, and the introduction of nonnative species, all of which alter historical conditions. Human interventions to restore ecosystem conditions are conceptualized as actions undertaken to re-create historical conditions. I suggest that if we are to understand novel ecosystems in which humans are the key players, we need a paradigm shift in the way we study these ecosystems.

In ecology, hybrid ecosystems represent the transition from a historical state to the novel ecosystem state on a two-dimensional scale defined by axes representing (1) the abiotic and (2) the biotic deviations from the historical condition. Hobbs, Higgs, and Harris (2009) described three system states: historical, hybrid, and novel (figure 2.3). They made a practical distinction between hybrid ecosystems, whose novelty it might still be possible to reverse, and novel ecosystems, which have crossed a threshold of irreversibility. Change may take many pathways; the transformations that drive it may be dominated by changes in biotic and/or abiotic elements and may cross thresholds that keep a system from reverting to its original state. In such a framework, the alternative system states are maintained by feedback mechanisms driven by the native or novel habitat.

To begin to understand the dynamics of novel coupled human-natural systems, we need to expand the dimensional space by adding a third axis: the evolution of the human habitat from Neolithic settlements to industrial and postindustrial cities (figure 2.4). As humans urbanize, novel biotic and abiotic conditions emerge. As hybrid ecosystems, cities fluctuate between two basins of attractions or alternative states, characterized by either historical or novel feedback mechanisms. As the planet becomes increasingly dominated by human action, such mechanisms

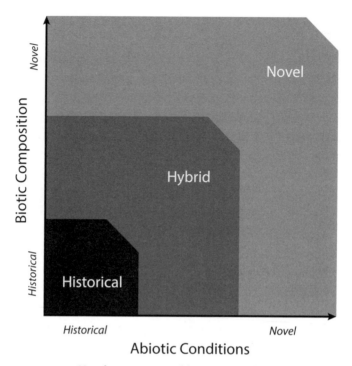

FIGURE 2.3 Novel ecosystem. Hobbs, Higgs, and Harris (2009) have described three system states: (1) *historical*, in which a system remains within its historical range of variability; (2) *hybrid*, in which biotic and abiotic conditions are dissimilar from their historical range but the system is able to return to its historical state; and (3) *novel*, in which the ecosystem is characterized by a regime shift.

move toward a new set of feedback mechanisms. Their state is unstable, not irreversible. Today different cities may occupy different positions within this multidimensional space. I suggest that most postindustrial cities are hybrid and unstable. We can drive them to unintentional collapse (so they revert to the historical, preindustrial state), or we can consciously steer them toward outcomes we desire (novel, resilient systems) (figure 2.5a–b).

The way to steer the city toward a resilient transformation is not yet clear, but cities across the world are investigating diverse avenues to do so. Cities present vast opportunities to explore alternative directions of

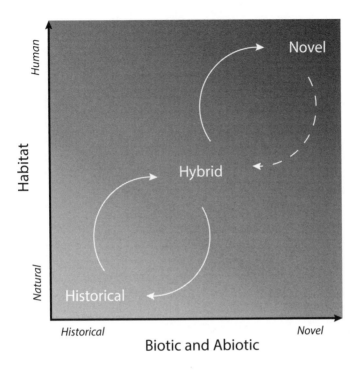

FIGURE 2.4 Coevolution of novel ecosystems: adding humans. Expanding the dimensional space of the ecosystem concept by adding a third axis allows us to view evolution of the human habitat from Neolithic settlements to industrial and postindustrial cities.

change and great potential to generate creative solutions. Increasing interactions between society and the environment are both the mechanism accelerating global environmental change and the source of innovation. In the following section, building upon complexity theory and resilience science, I articulate my hypothesis, and the chapter concludes by exposing implications and key scientific challenges for urban ecology.

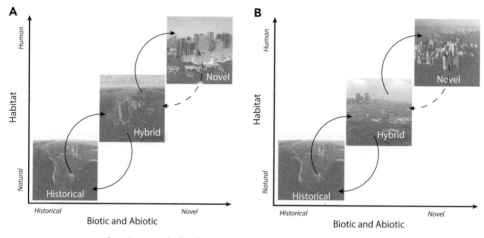

FIGURE 2.5a–b   Cities as hybrid ecosystems. As humans urbanize, (1) novel biotic and abiotic conditions emerge, or (2) humans consciously steer toward outcomes we desire (in a novel resilient system). Photographs: Manhattan (three hundred years' change): Eric W. Sanderson; Manhattan underwater: Cameron Davidson, CG imaging by John Blackford; Manhattan (Hurricane Sandy): Andrew Burton, Getty Images; novel Manhattan: Tingwei Xu and Xie Zhang.

The Structure, Dynamics, and Evolution of Hybrid Ecosystems

Hybrid ecosystems are characterized by complex interactions (e.g., heterogeneous, nonlinear, multiple scale), emergent properties (e.g., patterns, processes, feedback), multiple equilibria (e.g., regime shifts), and the capacity for innovation.

The emergence of cities is a prime example of complexity at work. Cities evolve through a complex series of interactions involving a vast number of different components, agents, and decisions. These components are themselves made up of smaller parts that exhibit their own dynamic behaviors. Yet at the higher scale of the unit hierarchy—neighborhood, city, or metropolitan region—new and unpredictable properties emerge. Complex systems can exhibit multiple configurations as a result of the evolutionary process (Wagner 2005). This quality allows the system to differentiate and still maintain preexisting functions. During such divergence, human and natural systems co-evolve, which ensures that they can maintain their functions despite the differences that they accumulate. Because of its potential for change and innovation, this co-evolution mechanism has a significant impact on the evolutionary process and the adaptive capacity of coupled human-natural systems.

Hybrid systems can experience completely novel interactions between human and ecological processes. In turn, such interactions produce novel ecological conditions and unprecedented expressions, leading to new ecological patterns, processes, and functions. An example is the interaction between seed dispersal and road transportation that takes place in urban environments. As humans become agents of seed dispersal, they facilitate competition among species, helping to determine which species thrive; meanwhile, the structures and infrastructure that humans build provide new vectors and pathways for seed dispersal and new habitats that determine seeds' survival. Vehicles have altered the mechanisms and patterns of seed dispersal in urban areas, making it far easier for nondispersing seeds to spread, sometimes farther than five miles (Von der Lippe et al. 2013). Other examples are the unintentional novel habitats of abandoned rail corridors and vacant lots.

Such novel interactions may also trigger unprecedented dynamics and unpredictable changes. Both ecosystems and societies experience continuous fluctuations in their structures and functions. Occasionally change is punctuated by a sharp shift: an abrupt transition to an alternative state with significant implications for system function and dynamics (Scheffer et al. 2001, 2012). Environmental change can result in changes that persist long after the removal of the external perturbation, due to a phenomenon called hysteresis—time lags in the responses of biological systems to environmental change—or the emergence of novel feedback mechanisms (Scheffer et al. 2001; Suding and Hobbs 2009). Emergent behaviors in complex systems may not have been seen or experienced in the past. Because complex systems are inherently unpredictable, uncertainty is a property of the evolution of a system.

At the same time, novelty in hybrid systems is a key component of reorganization and renewal. Urbanization modifies the spatial and temporal variability of resources, creates new disturbances, and generates novel competitive interactions among species. This fact is particularly important because the distribution of ecological functions within and across scales is key to a system's ability to regenerate and renew itself after the ecology of an area is disrupted. Peterson, Allen, and Holling (1998) pointed out that at different scales, species in a functional group mutually reinforce the resilience of a function and minimize competition among species within the group. They suggested that it is the presence of different functional groups within a scale and the replication of functions across scales that enable robust ecological functioning.

In hybrid ecosystems, innovation is an emergent property of complex coupling and feedbacks among human, natural, and technological system components (figure 2.5a–b, earlier). A more appropriate term to characterize innovation in such contexts is *socioecological innovation*— the emergence of novel *capabilities* emerging from new ideas, technologies, and formal and informal institutions that enhance human and ecosystem functions and their evolution. These encompass technological advances to fulfill human needs or purposes by rearranging matter, energy, and information, progress that both requires and enables social structure.

Although evolutionary biologists have recognized the significance of interactions in hybrids as a source of innovation in co-evolutionary processes, most researchers in ecology have seen the hybrid nature of urban ecosystems as a threat to ecosystem stability and resilience. In contrast, my hypothesis is that hybrid mechanisms are essential to maintaining ecosystem functions while simultaneously allowing systems to co-evolve and change. Understanding the bases of these newly generated interactions is central to understanding co-evolution and adaptation in hybrid systems.

Complexity

Cities are complex hybrid systems, not simply complicated ones. Complicated systems are predictable and controllable. We are often reminded of clocks, airplanes, and other machines that behave in fairly predictable ways—when they do not fail or crash. But in some ways, even crashes are predictable. On the other hand, complex systems are only partly knowable, highly unpredictable, and uncontrollable. These characteristics make city planning—both the study and practice—interesting and challenging. If cities were simply complicated systems, we would be able to predict their behavior with a reasonable degree of confidence, and we would have answers for almost all problems that arise. Cities are composed of multiple moving elements, diverse players, and independent mechanisms generating collective behaviors. Similar to other emergent phenomena, they are not controlled by a leader. Collective behaviors cannot be explained by the actions of individual agents taken separately. It is this complexity that makes cities evolve through growth and decline and face change by reconfiguring themselves in ways that their inhabitants could not have previously imagined.

But cities are not simply complex ecological or social systems. They are hybrids. The complexity of hybrid ecosystems has an enormous impact on the co-evolutionary process of coupled human-natural systems. Complexity provides hybrid systems with the ability to accumulate differences and evolve new functions while maintaining preexisting functions (Wagner 2005). In genetics, this is a fundamental source of novelty (Dover and Flavell 1984). Landry, Hartl, and Ranz (2007) described how novel phenotypes emerge from the interactions of two divergent genomes. Similarly, it is their hybrid nature that makes urban ecosystems capable of differentiating their functions while still maintaining the functions of their presettlement states.

For example, consider the case of delta cities and the emergent biodiversity of urbanizing estuaries. In such environments, barrier islands perform the hybrid function of protecting both biodiversity and human well-being (figure 2.6). Deltas are significant ecological resources for their diverse terrestrial and marine species. These areas are attractive to human settlement, too, as they are low-lying, highly productive, and rich in biodiversity and they offer easy transport along abundant waterways. Deltas cover 5 percent of the world's land area, but they are densely populated: over 500 million people live on them (Overeem and Syvitski 2009). Delta cities are particularly vulnerable to rising sea levels and increasing numbers and magnitudes of storm events, and the amount of land that is similarly vulnerable may increase by 50 percent, according to projections of sea-level rise, during the twenty-first century. For example, large portions, if not all, of New York, Miami, and Boston are expected to be submerged within the next few centuries (Strauss et al. 2012).

The biodiversity of estuaries in urban ecosystems emerges from interactions between ecosystem functions (e.g., natural habitat) and human functions (e.g., urban development and housing). Barrier islands are part

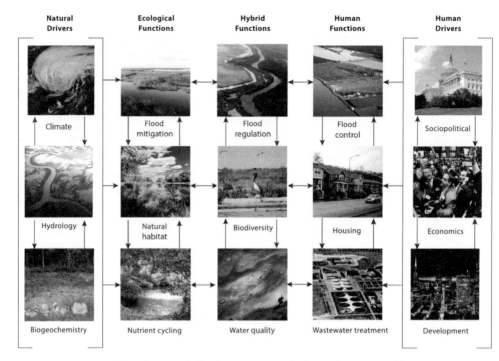

Natural Drivers	Ecological Functions	Hybrid Functions	Human Functions	Human Drivers
Climate	Flood mitigation	Flood regulation	Flood control	Sociopolitical
Hydrology	Natural habitat	Biodiversity	Housing	Economics
Biogeochemistry	Nutrient cycling	Water quality	Wastewater treatment	Development

FIGURE 2.6   Complexity in hybrid ecosystems, which are characterized by complex interactions (heterogeneous, nonlinear, multiple scale), emergent properties (e.g., patterns, processes, feedback), multiple equilibria (regime shifts), and innovative capacity. Photographs: climate: NASA Goddard Space Flight Center; hydrology: Feral Arts; biogeochemistry: Anders Sandberg; flood mitigation: Louisiana Travel; natural habitat: David Cornwell; nutrient cycling: Moni3 (Wikipedia); flood regulation: C. Stephen Haase/TNC; biodiversity: Coastal Wetlands Planning, Protection and Restoration Act; water quality: Chesapeake Bay Program; flood control: Tobin; housing: Joe Wolf; wastewater treatment: SA Water; sociopolitical: ttarasiuk; economics: thetaxhaven; development: Brandon Doran.

of a tightly coupled system of ecological drivers including tides, topography, sea level, and biogeochemistry; meanwhile, human drivers, including the construction of housing and infrastructure, are controlled by economic growth, urban development policies, and strategies for hazard management. New interactions and feedbacks emerge between urban patterns and socioecological processes, producing complex ecosystem dynamics (figure 2.7).

The newly heterogeneous types of landscape that urbanization pro-

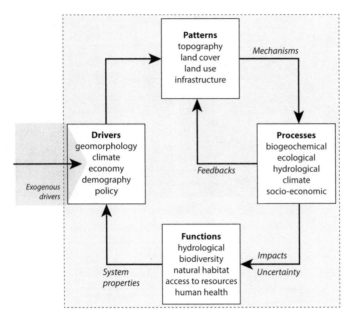

FIGURE 2.7   Urban ecology framework. In urban and urbanizing ecosystems, human and ecological functions and well-being are inextricably linked. Key human drivers of change are demographics, economics, technology, social organization, and political and governmental structures. People's choices about their location and their consumption behaviors directly influence the way they use land and demand resources. Urban patterns mediate relationships between drivers and human and ecosystem processes, and these affect urban ecosystem functions, which in turn influence the drivers of change. Source: Adapted from Alberti et al. 2003.

duces leads to complex patterns of runoff; meanwhile, levees and dikes cause trouble by trapping urban runoff water below river level, and local groundwater levels vary widely. Flood control strategies that depend on engineered built infrastructure interact with climate change to generate potential regime shifts such as flooding, loss of estuary biodiversity, fish kill, and algal blooms. Interactions between drivers occur across multiple temporal and spatial scales; they include global sea level rise and geologic formation, global trade and regulations, local-scale microclimates, point source pollution or microbial activity, and community practices and businesses.

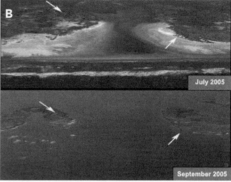

FIGURE 2.8a–b   (a) Emergent patterns. Novel hybrid patterns emerge from geomorphological processes and human intervention, as shown by channel migration
in the Mississippi River. Source: U.S. Geological Service in Julien and Vensel 2005.
(b) Emergent processes. Louisiana barrier island systems have experienced landward migration, area loss, erosion, and island narrowing. Source: U.S. Geological
Service in Julien and Vensel 2005.

Emergence

Ecosystem functions in urban ecosystems emerge from complex interactions between human and natural processes and are controlled by human
and natural drivers. Biodiversity and hydrological and biogeochemical
functions simultaneously support and are influenced by both human
and natural functions. In hybrid ecosystems, novel patterns, processes,
and feedbacks emerge. The amount and type of housing and patterns
of development are controlled by economic growth, demographics, and
policies. Similar to natural drivers, human drivers, such as economic
growth, sociopolitical forces, and development, interact with one
another, with strong feedbacks that shape human habitats.

In delta cities, for example, geomorphological processes and human
intervention combine to create novel hybrid patterns, as is shown by
channel migration in the Mississippi River (figure 2.8a, left). In parallel
with change in land cover over the past century, Louisiana's barrier island
systems have also experienced landward migration, loss of area, erosion,
and island narrowing as a result of complex interactions among subsid-
ence processes, sea level rise, storm impacts, inadequate sediment supply, and human disturbance, e.g., levees, canals, and seawalls (figure 2.8b,
right) (Julien and Vensel 2005). They also generate new feedbacks:

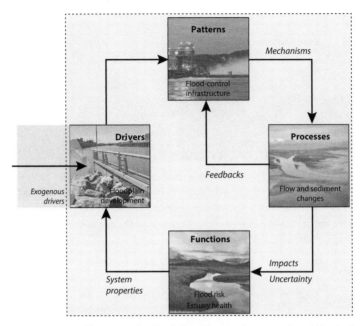

FIGURE 2.9   Emergent feedback. Land cover change and rapid loss of tidal marshes, coupled with the hydrological and ecological changes associated with the development of control structures (e.g., dikes, dams, levees, groins, seawalls, and storm water), make systems vulnerable to extreme climate events and prompt demand for flood-control infrastructure. Photographs: floodplain development: Lisa Campeau; flood-control infrastructure: Mark Lehigh; flow and sediment changes: Evan Leeson; flood risk and estuary health: DCSL.

changes in land cover and rapid loss of tidal marshes coupled with the hydrological and ecological changes associated with the development of hard control structures (e.g., dikes, dams, levees, groins, seawalls, and stormwater). All these make the system more vulnerable to extreme climate events and prompt more demands for flood control infrastructure.

As represented in figure 2.9, the flow and sediment alterations imposed by flood control structures (channelization, levees, and dams) prompt river adjustment—further geomorphic changes—that eventually affect flood risks by changing the structural integrity and capacity of flood control infrastructure. Flood development, as well as changes in flow and sediment, also causes direct biophysical changes that jeopardize the health of rivers.

FIGURE 2.10 Regimes in coupled human-natural systems. A system regime is maintained by mutually reinforced processes or feedbacks. Regime shifts are represented heuristically by a ball-and-cup diagram. The valleys or cups represent alternative regimes. The ball represents the state of the regime. Photograph: U.S. Army Corps of Engineers, Michael Maples.

FIGURE 2.11 Regime shift. A regime shift entails a shift in the current system state (represented as a ball) from one state to another. An external shock (such as Hurricane Katrina) can trigger a completely different system behavior. Photographs: left: U.S. Army Corps of Engineers, Michael Maples; right: aerial photo of Katrina, National Oceanic and Atmospheric Administration.

Regime Shifts

Interactions between human and natural drivers may generate regime shifts, such as flooding, loss of estuary biodiversity, and algal blooms. Regime shifts are shifts in dominant feedback. In urban ecosystems, they emerge from complex interactions between human and natural feedback mechanisms. Regime shifts are large, abrupt changes in the structure and function of a system, as illustrated here by the simple graph in figure 2.10. The term *regime* refers to the behavior of a system— in this case, a coupled human-natural system—that is maintained by

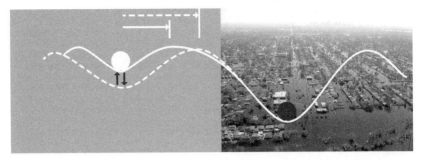

FIGURE 2.12   Regime shift. Slow changes in external drivers and/or internal feed-backs may reduce system resilience, illustrated here by the change in the width of the valley (or basin of attraction). The system moves from a resilient condition, represented by the dotted line, to a less resilient one, represented by the continuous line, changing the thresholds that push the system into a new regime. Photograph: aerial photo of Katrina, National Oceanic and Atmospheric Administration.

FIGURE 2.13   Regime shift. Loss of wetlands can increase vulnerability to extreme events (New Orleans). Photographs: estuarine marshes, Mississippi River Delta, John Barras, U.S. Geological Survey; aerial photo of Katrina: National Oceanic and Atmospheric Administration.

mutually reinforced processes or feedbacks. An external shock, such as a hurricane like Katrina, can trigger a completely different system behavior; in New Orleans, it is represented by a new hydrological regime (figure 2.11). But regime shifts also depend on slow changes in external drivers and/or internal feedbacks that alter the domains of attraction (or resilience) of the regime—from a resilient condition to a less resilient one—and change the thresholds that push the system into a new regime (figure 2.12). In New Orleans, for example, the loss of wetlands associated with land cover change and human intervention over the past fifty years has been a key reason that the urban ecosystem is now more vulnerable to extreme storm events (figure 2.13).

Innovation

The hybrid nature of urban ecosystems has been seen primarily as a threat to ecosystem stability and resilience. The higher concentration of people and increased social interactions in urbanizing regions are typically associated with increases in pollution, resource consumption, crime, and disease. Yet increased interactions among a diversity of people are what promote collaborative creativity and emergence of novelty. Bettencourt (2013) and his team predicted that the pace of social life in the city would increase with population size. As presented above, they have found ample evidence across diverse world urban regions that many emergent properties of cities, from patent production and personal income to electrical cable length across thousands of cities worldwide, are power-law functions of population size, with scaling exponents that fall into distinct universality classes.

At the core of innovation in ecosystems is the fact that species are often not continuous across scales. Discontinuities in ecosystems emerge when adaptive cycles are separated by gaps in the domains of scale that they occupy; these gaps often span at least one order of magnitude (Gunderson and Holling 2002). By linking species interactions to ecosystem functions across multiple scales, Peterson, Allen, and Holling (1998) hypothesized that cross-scale functional arrangements provide a robust response to a variety of perturbations such as insect outbreaks. Distinct scale breaks emerge, reflecting abrupt changes in pattern and structure that create discontinuities in scale. In turn, these provide opportunities to experiment within a given scale and to develop high levels of diversity across scale (O'Neill et al. 1986). It is the inherent variability present in cross-scale structures that enables innovation and transformation (C. R. Allen and Holling 2010).

The capacity for innovation is an essential precondition for a system to function. Gunderson and Holling (2002) have noted that without innovation capacity and novelty, systems may become overconnected and dynamically locked, unable to adapt. To be resilient and evolve, they must create new structures and undergo dynamic change. Differentiation, modularity, and decentralization of organizational structures have been described as key characteristics of systems that are capable of simultaneously adapting and innovating (Pierce and Delbecq 1977). As we learn more about the complexity of the mechanisms that regulate coupled human-natural systems, we can better explore the links between discontinuities in hybrid ecosystems and the novel expressions that enable them to evolve.

In hybrid ecosystems, cross-scale interactions between human and natural systems have the power to generate completely new configurations. Under unstable critical transitions, increasing social interactions can generate *novel actions*. For example, as a result of critical transitions during hurricanes Sandy and Katrina, people developed new coalitions, collaborations, and creative strategies, reconfiguring system responses. They also developed other novel actions, relying on real-time communication, cell phone networks, online delivery systems, and the redundancy and flexibility of multiple transportation and infrastructure systems (Ernstson et al. 2010).

The evolutionary process leads hybrid systems to exhibit multiple configurations: they retain elements from historical ecosystems while simultaneously introducing new structures and dynamics. Ecological memory—the capacity of a past ecosystem state to determine its future trajectory (Padisák 1992)—consists of both biological legacies (e.g., remnants of species) and genetic legacies (e.g., genetic diversity) that persist after a perturbation. If the ecosystem is to recover after a collapse, it needs to retain elements of the historical system in its ecological memory, to draw on past events that can inform its present or future responses (Peterson 2002). Meanwhile, new elements enhance evolution of the human habitat.

Critical Transitions, Resilience, and Transformation

Ecological resilience, as defined by resilience theory, corresponds to the width of a system's stability landscape—that is, its ability to fluctuate and adapt. Resilience is the capacity of a social-ecological system to absorb shocks or perturbations and still retain its fundamental function and structure (Holling 1973) (figure 2.14). From a system dynamics perspective, as a region urbanizes, the ecological processes supporting the urban ecosystem may reach a threshold and drive the system into an unstable

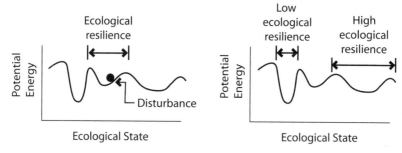

FIGURE 2.14 Ecological resilience, a measure of the amount of change required to transform the set of mutually reinforcing processes and structures that maintain a system, as represented by the width of its stability pit. Source: Adapted from Home Planet, http://home.planet.nl/~loon0781/fase1.pdf.

state (figure 2.15). Eventually the system shifts into a new state in which ecological processes are highly compromised (Alberti and Marzluff 2004). Alternatively, I hypothesize, a novel ecosystem state evolves (figure 2.16).

The concept of adaptive co-evolution provides a useful framework to explore the dynamics of urban ecosystems. Adaptive evolution occurs largely as minor variations in phenotype successively accumulate; this accumulation eventually leads to changes in the environment and, ultimately, to evolutionary change in other organisms (Kauffman and Johnsen 1991; Kauffman 1995). The concept of "fitness landscape"—the idea that species evolve to fit the landscape around them—means that species and landscape evolve together. Kauffman (ibid.) noted that through this process, species change the very mechanisms through which they co-evolve. Humans challenge the cultural and genetic makeup of species and their ecosystem functions by affecting the fitness landscape in which they evolve. Humans thus change opportunities for the co-evolution of human-natural interactions.

What makes an urban ecosystem simultaneously resilient and able to change? Do urban ecosystems have generic properties or qualities that predict their adaptive capacity? By combining emerging insights from diverse research domains, Scheffer et al. (2012) conjectured that the nature of interactions among system components is one key to understanding how the properties of systems affect system stability and resilience. Scientists studying complex systems show that highly heterogeneous

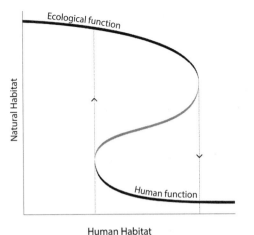

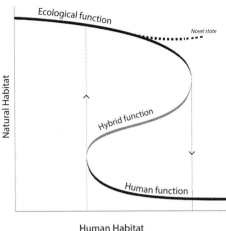

FIGURE 2.15   Resilience in hybrid ecosystems. Resilience is the capacity of a social-ecological system to absorb shocks or perturbations and still retain its fundamental function and structure.

FIGURE 2.16   Novel state in hybrid ecosystems. In hybrid ecosystems, the emergence of hybrid functions might allow a shift into a novel ecosystem state.

systems with modular structures tend to experience greater resilience and adaptability to changes than do those that are highly connected and homogeneous. Emerging evidence from the study of coupled human-natural systems also indicates some properties of complex systems that enhance adaptive capacity and innovation; among them are cross-scale interaction, early warning, and self-organization (Walker et al. 2004) (see chapters 4 and 10).

A notion that begins to explain scale-free phenomena in natural systems is self-organized criticality: systems spontaneously evolve toward a critical point (Solé et al. 1999). The best-known example is provided by Bak (1996), who observed the behavior of a sand pile. Imagine pouring sand continuously onto a pile. The sand accumulates in height until the pile reaches a height-to-base ratio threshold that triggers an avalanche. Bak, Tang, and Wiesenfeld (1987) advanced that avalanches have different sizes and follow a power-law distribution. Initially most of the avalanches are small, but the range of sizes grows with time until the pile reaches a critical state. Physicists have observed that systems near a phase transition become critical: they do not exhibit any characteristic scale, and they spontaneously organize themselves in fractals. Critical systems behave in a simple manner: they obey a series of power laws (Gisiger 2001).

Questions that intrigue ecologists are whether ecosystems are critical and how criticality relates to their evolution (Solé et al. 1999). In studying genetic networks in biological systems, Torres-Sosa, Huang, and Aldana (2012) found that critical systems exhibit important properties—fast information processing, collective responses to perturbations, and the ability to integrate a wide range of external stimuli without saturation—that allow robustness and flexibility. In biological systems, criticality is a fundamental mechanism that generates dynamic robustness for the network and meanwhile allows the network to respond to perturbations.

Several authors have observed that in these tightly coupled systems, resilience and adaptation emerge from the inherent uncertainty of complex cross-scale human-environment interactions, which are variable in both space and time (Brooks 1986; Holling 1995). A key factor governing such interactions may be the lag time between human decisions and their impacts, delayed and distributed over long distances (Alberti 2007). Such interactions in hybrid systems are highly influenced by technology. Other authors have pointed out that public infrastructure is a key variable affecting both the ways that these systems can function and their resilience (Anderies, Janssen, and Ostrom 2004) because it regulates the relationships between humans and natural resources through both physical and social mechanisms (Costanza et al. 2001). In turn, both the physical and social capital that characterize public infrastructure systems and the diverse institutions that govern them regulate strategic interactions and feedbacks.

Interactions both drive and are driven by change. Environmental and socioeconomic changes simultaneously affect socioeconomic systems and the infrastructures that support them. Meanwhile, changes in internal feedback mechanisms (e.g., reduction in resilience) can cause the larger system to reorganize rapidly. Thus, efforts to characterize how a human-dominated ecosystem responds to change and its potential regime shifts will require anticipating thresholds by understanding the mechanisms and feedback loops that govern the ecosystem's functioning.

An important shift in resilience science has emerged from the study of cities as coupled human-natural systems (Pickett, Cadenasso, and

McGrath 2013). Such theoretical insights have begun to inform studies of human responses to climate change and the designs of adaptation strategies. In ecosystem studies, humans have been represented as external elements, both as subjects of inquiry and as agents of inquiry. In coupled human-natural systems, humans are endogenous variables and agents of knowledge production. Human-natural interactions change as the system evolves, and they acquire new information through a series of feedback mechanisms (Anderies, Jansson, and Ostrom 2004). Remarkably, one piece of evidence supporting the concept that human-natural interactions can learn and evolve is the way that human settlements have evolved in response to socioeconomic and environmental change.

Can the emerging patterns of urban agglomeration affect the probability that we will cross thresholds that will trigger abrupt changes on a planetary scale? The number of megacities has nearly tripled over the past two and half decades, and megaregions of urban agglomeration are arising all over the world (United Nations 2014). Resulting social, technological, and economic growth is accompanied by human-natural interactions of unprecedented scale (Ross 2009; McHale et al. 2015). As structures and functions of cities evolve, so too do their interactions with the environment (Pickett and Zhou 2015). These shifts potentially could generate critical transitions on a planetary scale (box 2.1).

Box 2.1. Can Emerging Urban Agglomerations Cause a Tipping Point?

Over the past six decades, our planet has undergone rapid urbanization. In 1950, less than 30 percent of the world's population lived in urban settlements (United Nations 2014). By 2014, that figure had risen to 54 percent, and by 2050 it is expected to be 66 percent. Population growth and urbanization are projected to add 2.5 billion people to the world's urban population, with 90 percent of the increase concentrated in Asia and Africa. Although nearly half of urban dwellers reside in settlements of fewer than five hundred thousand people, about one in eight (12.5 percent) live in one of twenty-eight megacities, those with populations exceeding 10 million (UN-Habitat 2008) (figure B2.1). The

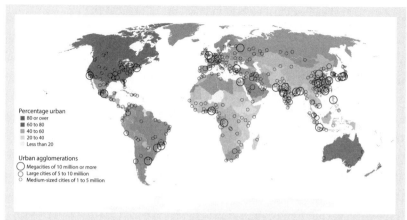

Percentage urban
■ 80 or over
■ 60 to 80
■ 40 to 60
■ 20 to 40
   Less than 20

Urban agglomerations
◯ Megacities of 10 million or more
◌ Large cities of 5 to 10 million
○ Medium-sized cities of 1 to 5 million

FIGURE B2.1 Emerging urban agglomeration patterns (United Nations 2011).

number of megacities has nearly tripled since 1990, and by 2030, the planet will house forty-one of these urban agglomerations.

Rapid urbanization poses significant challenges for both human and ecological well-being. Most urban growth is occurring in emerging economies, where rapid rural-to-urban migration generates significant shifts in socioeconomic systems. Such transitions, associated with poorly planned settlements, have driven the emergence of informal settlements such as slums—densely populated urban areas characterized by crowded and poor-quality housing and by inadequate public services and access to infrastructure (Marx, Stoker, and Suri 2013). High population density, in the absence of appropriate sanitation and basic services, has profound impacts on human and environmental systems (United Nations 2011; Glaeser 2012). Many urbanizing regions cannot keep pace with the infrastructure needs of their growing populations. Estimates suggest that more than a billion people live in informal settlements and that about 40 percent of the world's urban expansion is taking place in slums, exacerbating socioeconomic disparities and environmental deterioration (United Nations 2013).

Urbanization is driving systemic changes to socioecological systems associated with globalization by accelerating rates of interactions among people and places, multiplying numbers and strengths of connections, expanding the spatial scales and influences of human activities to global levels, and changing biotic and social heterogeneity (Young et al. 2006). Increasing evidence shows that cities accelerate the time-space compression associated with globalization (Harvey 1989) and expand the consequences and interactions of interrelated phenomena such as the movements of people and products, access to and disruption of natural resources, and threats to biodiversity (Lenzen et al. 2012). These distant interactions pose simultaneous challenges and opportunities for sustainability (Liu et al. 2013).

Together with urban population growth, accelerating socio-economic and technological changes have driven a new configuration of urban settlements, redefining the concepts of cities and regions and dissolving the boundaries among urban, suburban, and rural areas (UN-Habitat 2008). The emerging urban agglomerations extend to include urban centers of varying density, along with suburban areas, commercial and business nodes, transportation corridors, industrial parks, and fragments of wild and rural lands (Ross 2009; McHale et al. 2015). These agglomerations span political borders (e.g., the Boston–New York–Washington corridor and the Shanghai–Nanjing–Hangzhou triangle) and act as major engines of the global economy (Sassen 2012).

Evidence of the emerging new spatial order first appeared more than half a century ago. French economic geographer Jean Gottmann (1961: 9) referred to it as "the dawn of a new stage in human civilization." Gottmann used the term "megalopolis" to describe the convergence of cities on the U.S. Eastern Seaboard, from Boston to Washington, DC. Gottmann described the conurbation, which then stretched for 600 miles and included 30 million people, and saw the Eastern Seaboard megalopolis as an experiment that would apply to other parts of the country by the end of the 20th century (ibid.). Today convergences of megalopolises are defined as *megaregions*: networks of metropolises characterized by multinodal mosaics of developed and undeveloped lands (Harrison and Hoyler 2014; Ross 2009).

There are forty megaregions in the world. They house 1.2 billion people—18 percent of the global population—and account for about 66 percent of the world's economic activity. In the United States, there are ten megaregions, accounting for 30 percent of national territory and 75 percent of the nation's population and employment. Today the

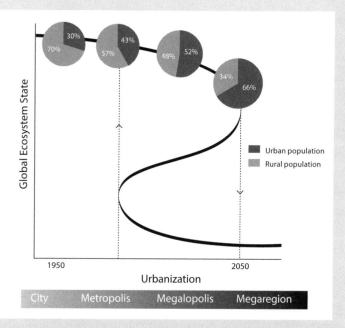

Global Ecosystem State

Urban population
Rural population

1950   2050

Urbanization

City   Metropolis   Megalopolis   Megaregion

FIGURE B2.2   Emergent patterns as tipping point. Modified from Barnosky et al. 2012. Population projections: UN 2014.

Boston–Washington megaregion houses 18 percent of the United States' population—over 56 million people—and its annual output is $3.75 trillion, or 23 percent of the U.S. GDP (Florida 2013). By the year 2050, two-thirds of the U.S. population will live in megaregions.

Can emerging patterns of urban agglomeration affect the probability of crossing thresholds that will trigger abrupt changes on a planetary scale (figure B2.2)? Urban regions have emerged as major drivers of environmental change. If all areas with high probabilities of urban expansion undergo this pattern of change, then by 2030, urban land cover will increase by 1.2 million square kilometers, nearly tripling the global urban land area of 2000 and resulting in massive habitat loss in key biodiversity hot spots.

To articulate this hypothesis, we must examine key structural and functional shifts associated with the emergence of urban agglomerations. The dominant city structure has evolved over the past century from the centralized settlement of a region to a polycentric structure by increasing its connections to satellite cities of the rural hinterland, eventually giving rise to the metropolis. This set the stage, in turn, for the development of more diffuse, multicentric structures—megacities—and, finally, for the emergence of the modern structure: networked megaregions (Pickett and Zhou 2015).

This evolution is a result of changes in functions. What were chiefly industrial centers have become economically diverse, as service- and knowledge-based economies have emerged, regional and global integrations have arisen, and shifts in infrastructure and technology have driven the rapid formation of teleconnections. Urban agglomerations have become more connected and interdependent, shifting the focus of key linkages from rural-urban to regional and global (Gottmann 1961; Hall 2001; Vicino et al. 2007; Ross 2009; McGrath and Shane 2012; McHale et al. 2015).

These shifts potentially could generate critical transitions in social-ecological dynamics and in global ecosystem states by amplifying environmental change through several mechanisms, including land transformation and associated changes in habitat and biotic interactions, increased magnitudes and frequencies of disturbances, and the emergence of novel disturbances (Pickett and Zhou 2015). The increased social and economic interactions of emerging megaregions expand the edges of urban socioecological dynamics beyond city boundaries and accelerate changes at multiple scales. Megaregions exacerbate tele-coupling: interactions between distant places that often lead to unexpected outcomes (Liu et al. 2013).

At the same time, however, these amplifications of cross-scale interactions provide new opportunities for both technological and social innovation. Increased connections among people and places that occur across scales promote faster communication and learning. Parallel to structural and functional changes, emerging megaregions represent a shift in governance from formal centralized systems to hybrid multiscale, cross-regional network structures that include both formal and informal institutions. The emergence of new interactions and feedbacks across distant places and between local and global processes in megaregions offers unique opportunities to experiment with novel institutional frameworks that facilitate the discovery of novel solutions to complex problems. ●

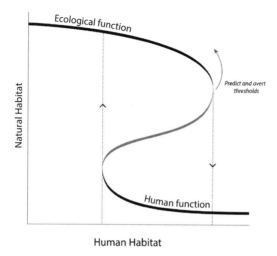

FIGURE 2.17 Thresholds for predicting and averting regime shifts in urban ecosystems.

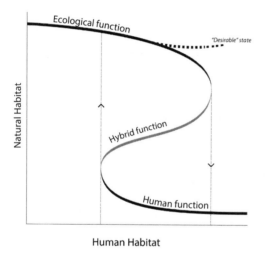

FIGURE 2.18 Desirable states in hybrid ecosystems.

Toward a Co-Evolutionary Perspective

Resilience theory offers a robust framework for understanding what triggers regime shifts and the alternative pathways through which ecosystems and societies navigate change. But is persistence always desirable? Many undesirable states, such as degraded ecosystems and social inequalities, may be quite resilient. So while resilience planning focuses on predicting and averting potential overrun of thresholds and system shifts (figure 2.17), actually moving toward a desirable state may require trans-

Table 2.1. Planning paradigms, simplified representation

Characteristic	Steady State*	Resilience	Co-evolution
Reference	Historic condition	Historic range of variability	Trajectory of change
World view	Stability	Multiple equilibria	Shifting stability
Treatment of uncertainty	Minimize uncertainty	Incorporate uncertainty into decision-making	Maximize flexibility to adapt to uncertain future condition
Strategic goal	Minimize impact and disturbance	Maintain variability and redundancy	Maintain self-organization
Policy approach	Optimization	Adaptation	Transformation

*Steady state is a view in which nature exists at or near an equilibrium condition. In this paradigm, disturbance can be controlled, and optimization is the strategy to achieve sustainable carrying capacities. The resilience paradigm recognizes the existence of multistable states and focuses on adaptation as a strategy to maintain system function. A co-evolutionary perspective of human-natural systems encompasses a shift in system stability that might require transformation toward what is desirable, which implies addressing the diversity of human values. It also implies redefining the reference conditions and requires institutional flexibility, learning, and reconfiguration through active policy.

Table is modified from Chapin et al. 2010.

formation, which implies a shift toward a novel state (figure 2.18), not simply the adaptation of both human and natural systems.

A co-evolutionary perspective may be more appropriate to address the tension between resilience and transformation. It is a view that focuses not only on unpredictable dynamics in ecosystems, but also on institutional and political flexibility for learning, reconfiguration, and active policy. Coupled human-natural systems require a shift in both science and planning paradigms (table 2.1). While the resilience paradigm has offered a useful alternative to a steady-state paradigm, it is incomplete: it recognizes the existence of multiple equilibria and focuses on adaptation as a strategy for maintaining system function, while co-evolution in coupled human-natural systems implies both resilience and transformation (Folke 2006; Chapin et al. 2011).

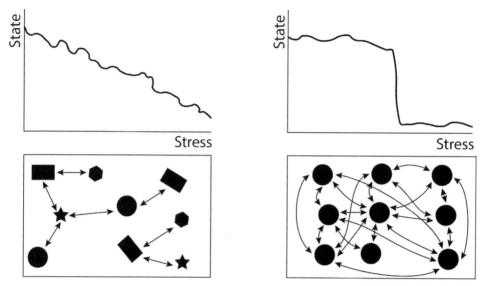

FIGURE 2.19   System characteristics. Emerging evidence indicates that systems with greater heterogeneity (i.e., greater variation among components) and modularity (i.e., incomplete connectivity) tend to have greater adaptive capacity than those whose elements are highly connected and homogeneous (Scheffer et al. 2012).

We can build on the emergent theory of critical transitions, resilience, and transformation to develop a science of cities as hybrid ecosystems. Emerging evidence indicates that systems that are more heterogeneous (in which the components vary) and modular (not completely connected) tend to be better able to adapt and evolve than those with highly connected homogeneous elements (figure 2.19). Building on these findings, I propose that we develop and test the hypothesis that the diversity and modularity of hybrid urban ecosystems allow us to promote transitions toward novel, resilient systems. Furthermore, I hypothesize that cross-scale interactions and discontinuities point to ways in which systems can change and evolve (or co-evolve) and provide us with opportunities as well as challenges (C. R. Allen et al. 2014) (figure 2.20).

To study resilience and transformation in urban ecosystems, we not only need to identify the generic properties or qualities of their system architecture that allow these ecosystems to persist and change; we also—and especially—need to establish a series of causal links and feedbacks among several factors: drivers of urbanization, patterns of

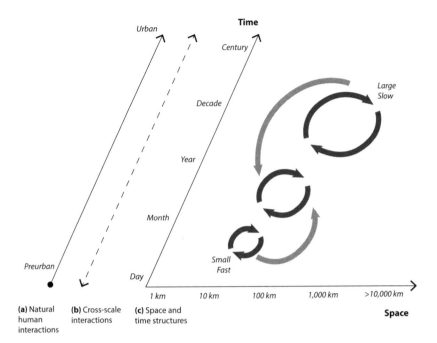

Time

Urban

Century

Decade

Year

Month

Small
Fast

Preurban

Day

Large
Slow

1 km   10 km   100 km   1,000 km   >10,000 km

(a) Natural
human
interactions

(b) Cross-scale
interactions

(c) Space and
time structures

Space

FIGURE 2.20   Cross-scale interactions. The diagram represents the adaptive cycle in a multidimensional space of space and time. Increasing cross-system interactions between human and natural systems, from preurban to more human-dominated systems, increases instability. It simultaneously provides opportunity for innovation.

resource use and land use, infrastructures driven by urbanization, and both biophysical and socioeconomic processes and responses (see figure 2.7) (Alberti 2008). And we must assess how these affect ecosystem function at multiple scales of time and space and account for potential trade-offs (Chelleri et al. 2015).

Scientific Challenges

In conclusion, I highlight some scientific challenges that hybrid, coupled human-natural systems pose to urban ecology and resilience science. Chiefly, these are the challenges associated with designing robust studies to uncover a diversity of mechanisms governing urban ecosystem dynamics (figure 2.21).

Complexity   Scales Mismatch   Uncertainty   Measurements   Assessment

FIGURE 2.21   Scientific challenges (five images). Images are iconic representations of the five challenges. Complexity, image: Grant Dawson. Scale, graphic: Jingle UW 2009.

Complexity

The first challenge that scientists face in studying urban ecosystems is the complexity of interactions between natural and human processes, which also involve a diversity of agents and institutions. This complexity has implications for the nature of empirical tests in urban ecological study and the ability of urban ecology scholars to establish empirical evidence of the many relationships that they hypothesize exist between human and ecological systems. The presence of multiple confounders challenges their ability to establish causality. Contradictory findings illustrate the challenges of defining and assessing ecological and human well-being in urban ecosystems. What is the evidence behind the claims that trees in cities are beneficial for people and ecosystems everywhere, regardless of culture, climate, governance, and infrastructure (Pataki et al. forthcoming)? Some scholars question whether unmeasured confounders may be operating (Frumkin 2013). Furthermore, most of the definitions and classifications used to characterize nonurban environments may not be appropriate in human-dominated systems. We must refine the way we characterize human-natural interactions by explicitly representing urban gradients, patterns, processes, mechanisms, effects, legacies, and scales.

Scale

Human and natural processes operate at different and variable scales, leading to mismatches of scale. These mismatches pose significant challenges for research design: At what scale should we study coupled human systems, and how do we define system boundaries and feedbacks? Scale mismatches between ecological processes and the institutions that govern natural resources may also play a key role in reducing resilience, prompting new questions about the scales appropriate for designing institutions to manage social-ecological systems (Cumming, Cumming, and Redman 2006). To answer these questions, we need to conduct studies that explicitly represent multiple spatial and temporal scales and investigate cross-scale interactions and effects.

Uncertainty

Uncertainty and surprise in urban ecosystems are amplified by the complex interactions among ecological and social drivers and their unpredictable dynamics. Traditional approaches to planning and management typically rely on predictions of probable futures extrapolated from past trends. However, predictive models that are designed to provide accurate assessments of future conditions can account for only some interactions between highly uncertain drivers of change and surprising, but plausible, futures over the long term. Despite remarkable progress in complex modeling and improved simulation and computer power, our models are still constrained by our limited knowledge, unverified assumptions, and short-sighted mindsets. We need to learn to embrace the lack of predictability and certainty inherent in hybrid systems and to expand our toolbox for exploring the future by integrating observation, models, and plausible scenarios to test hypotheses of human and natural interactions and systematically assess impacts and tradeoffs.

Measurements

A key challenge is robust measurement. Most assumptions regarding the mechanisms governing pristine environments do not hold true in urban settings. But the methods of observing and measuring applied in urban ecosystems still reflect such assumptions. Humans in urbanizing regions affect ecosystem processes through distinctive factors (e.g., the extent of

Box 2.2. Defining *Reference Condition* in Hybrid Ecosystems: What Can We Learn from Venice?

The city of Venice, Italy, is an emblematic example of how human settlements evolve in response to environmental change. Founded in the fifth century on a collection of 118 small islands, Venice had become a major maritime power by the tenth century. During the millennium and more since then, Venice has directed its own evolution, shaping its lagoon through human interventions that interact closely with the natural environment. Doing so has required extraordinary levels of human creativity and technological innovation—Venice has seen one of the world's greatest experiments in constructing city infrastructure in response to environmental change.

But the vast transformation of natural processes in Venice's lagoon is also representative of the irreversible processes at play in coastal cities in general. Driven by the global rise in sea levels over the past century, the mean water level around Venice has risen more than three

impervious surfaces, the built infrastructure, demographic trends, and human preferences) that simultaneously affect biophysical processes, nutrient cycling, and species competition. For example, assumptions regarding allometric models relating tree cover and carbon sequestration do not hold true in urban settings because these models were developed in forest ecosystems.

The challenges in urban ecology lie in both the lack of validated measurement methods and the limited availability of cross-site longitudinal data that can be used to assess ecosystem function along gradients of urbanization across different regions. Further challenges lie in efforts to quantify human well-being. Several methodological questions regarding dose responses, biological mechanisms, and confounders challenge the epidemiological foundations of the claim that trees benefit human health (Frumkin 2013). Should tree planting be encouraged in all cities, regardless of location, species, and cost of maintaining the trees? To inform urban planning and design, we must develop, validate, and inches since 1897. Meanwhile, the city's land has dropped by six inches, accelerated by groundwater pumping for industrial use, so it now experiences flooding events several times per year. The environmental changes in the lagoon are amplified by increasing erosion and disappearance of salt marshes and mud flats (Deheyn and Shaffer 2007). Human and natural functions of the urban ecosystem in the Venetian lagoon cannot be maintained without creative human intervention.

The MOSE Project (Modulo Sperimentale Elettromeccanico; www .mosevenezia.eu/) is one of the more recent attempts to control flooding and prevent Venice from further sinking. A complex $6.5 billion system, it attempts to mimic natural processes by creating adjustable barriers at entrances to the lagoon. Seventy-eight mobile and articulated barriers at the three lagoon inlets lie on the sea floor during normal sea and weather conditions, allowing natural exchanges between the lagoon environment and the Adriatic Sea (figures B2.3– B2.4). During high seas and other times when weather conditions could lead to flooding, the barriers rise to their upright positions,

monitor comparable metrics across cities, regions, and biomes so we can assess mechanisms under different development patterns.

Reference Conditions

Finally, how can we define reference conditions for resilience and transformation in coupled human-natural systems (box 2.2)? If the historical condition does not apply when evaluating the functioning of a novel ecosystem, which reference criteria can be applied in its place? How can a co-evolutionary paradigm integrate ecology and human functions? What is the role of human values? Coupled human-natural systems require transformation toward what is desirable, and that requirement implies addressing the diversity of human values and the existence of conflicts. Such questions demand a new ethical framework and pose important questions about the relationships among environmental ethics, natural history, and science.

FIGURE B2.3 Satellite image of the Venice lagoon and the three inlets of Lido, Malamocco, and Chioggia, where the Modulo Sperimentale Elettromeccanico (MOSE) antiflooding project is being constructed, 2009. Source: Magistrato alle Acque di Venezia—Consorzio Venezia Nuova, 2009.

Venice

Lido Lagoon Inlet

Malamocco Lagoon Inlet

Chioggia   Chioggia Lagoon Inlet

FIGURE B2.4 MOSE project, Lido inlet, 2009. Source: Magistrato alle Acque di Venezia—Consorzio Venezia Nuova.

preventing inundation. But the development and implementation of the project is not without controversy. At first, environmentalists were concerned about its impacts on the delicate lagoon ecosystem. Today the major concern is whether this infrastructure will even be able to withstand anticipated climate change.

MOSE aims to create a resilient Venice, but its design is not intended to re-establish the lagoon's natural hydrological regime based on conditions before the city was settled. A major challenge for urban ecologists is to define reference conditions. Can presettlement natural conditions serve as a suitable reference for a functional coupled human-natural system? Probably not (Marani 2013). Interaction between coastal human settlements and ecosystem processes leads inevitably to the emergence of new dynamics and feedbacks.

Cities such as Venice, as they work to adapt to current and future climate impacts, must look beyond their presettlement natural state for their reference conditions, and perhaps even beyond any situation they have encountered in the past. To define a reference condition for a given hybrid ecosystem may require a new kind of thinking: a process guided by both science and imagination, in addition to observations and extrapolation from the past. ●

3

Reframing Urban Ecology

IN THIS CHAPTER, I EXAMINE THE ORIGIN OF URBAN ECOLOGY AND its evolution toward a science of coupled human-natural systems. During the past hundred years, advances in the scientific understanding of ecological systems have opened up new opportunities to integrate humans into ecology. Yet despite increasing human domination of Earth's ecosystems, ecological theories are still conceptualized primarily in terms of biophysical, ecological, and evolutionary processes that are substantially unaffected by human agency. I discuss observations of "anomalies" in urban ecosystems that are difficult to reconcile within ecological theory and the emergence of a co-evolutionary paradigm that better explains patterns observed in urbanizing regions. I conclude the chapter by highlighting emerging elements for reframing the science and practice of urban ecology.

Ecology for an Urban Planet

The science of ecology has evolved through many stages and definitions (Likens 1992; O'Neill 2001; Marquet et al. 2014). What characterizes its evolution is the dynamic interplay between changes that have occurred in Earth's ecosystems over time and parallel shifts in our scientific understanding. It is through such interplay, over the past century, that new ideas have developed (Graham and Dayton 2002). The emergence and evolution of the ecosystem concept provided a framework for studying ecological interactions among individuals, populations, and communities and their abiotic environments, and for studying the changes in these relationships over time (Likens 1992). Although the term *ecosystem* was coined by Tansley in 1935 and the underlying idea can be traced back to Marsh (1864), the concept of *ecosystem* became a standard paradigm for studying ecological systems only when systems analysis emerged in the second half of the past century (Holling 1973; Odum 1971). Since then, ecological scholars have revised the concept to acknowledge *multiple equilibria* and the *open, hierarchical, spatially heterogeneous*, and *scaled* nature of ecosystems (Levin 1999; O'Neill et al. 1982; Pickett, Parker, and Fiedler 1992). Still, the term essentially excludes humans.

The *ecosystem concept*, which is at the core of ecological thinking, emerged from a view of systems theory that is difficult to reconcile with our current understanding of dynamic ecological systems that may operate far from equilibrium (O'Neill 2001, 3275). In his MacArthur Lecture in Ecology, Robert O'Neill (ibid., 3276) proposed that in ecology the "ecosystem concept" has constituted a paradigm in the Kuhnian sense, an "*a priori* intellectual structure" rather than an "*a posteriori* empirical observation" about how nature works. While the concept offers a practical approach to the study of ecological systems—one that has proved instrumental in ecology's progress—it has some drawbacks. O'Neill argued powerfully that the ecosystem paradigm accepts a set of assumptions that limits our thinking and the questions we ask because it emphasizes some properties of nature while ignoring others. For example, emphasis on the self-regulating nature of ecosystems has biased ecosystem scientists to see disturbance as interference from outside. Yet we know that disturbance regimes are critical to understanding stability and ecosystem function.

A key property of ecosystems is their ability to change state in response to a spectrum of variable conditions (Holling 1973). Ecosystems have evolved over millions of years through changes in biotic-abiotic interactions. Since the Industrial Revolution, humans have increasingly dominated such interactions. But in ecology, humans are the only species that is considered to be external to ecosystems. They are seen as consumers of ecosystem services rather than active participants in ecosystem processes. Yet today humans are creating novel ecosystem functions well outside the range of values and conditions that Earth's ecosystems have experienced throughout their evolution (O'Neill 2001; Tilman and Lehman 2001).

If ecology is the study of organisms' interactions with one another and with the environment, and of the transformations of matter, energy, and information, then the advent of the Anthropocene implies a fundamental shift in ecology as a science. Earth's atmosphere, on which we all depend, emerged from the metabolic processes of vast numbers of single-celled algae and bacteria living in the seas 2.3 billion years ago. These organisms transformed the environment into a place where human life could develop. From a planetary perspective, the emergence and rapid expansion of cities across the globe may be a turning point in the life of our planet, one on the scale of the Great Oxidation. Recent increases in positive feedback (e.g., climate change), along with the emergence of evolutionary innovations (e.g., novel metabolisms) associated with the Anthropocene, could trigger transformations of the same magnitude and significance (Lenton and Williams 2013).

During the past hundred years, advances in the scientific understanding of ecological systems have created new opportunities to integrate humans into ecology (Alberti et al. 2003). Evolutionary theory and population genetics have fundamentally changed assumptions underlying ecological research. Ecological scholars no longer regard ecosystems as closed, self-regulating entities that "mature" to reach equilibrium. Instead, they see such systems as having multiple equilibria and being open, dynamic, highly unpredictable, and subject to frequent disturbance (Pickett, Parker, and Fiedler 1992). In the new, nonequilibrium paradigm, succession has multiple causes, can follow multiple pathways, and is highly dependent on environmental and historical context. Ecosystems are driven by processes (rather than end points) and are often regulated by external forces (rather than internal mechanisms).

In the second part of the past century, the new ecological paradigm led to an explicit acknowledgment that humans are components of ecosystems—and to the emergence of contemporary urban ecology (McDonnell and Pickett 1993; Pickett, Cadenasso, and McGrath 2013; McPhearson et al. 2016). In prior ecological theory, humans were considered to be external agents of disturbance (B. L. Turner and Meyer 1993), and, despite new attention to human-dominated habitats, initial ecological studies of cities still reflected that earlier perspective. For example, they focused on ecological phenomena such as the patterns and mechanisms controlling the functioning of plant communities and their changes (Cadenasso and Pickett 2013). Both the initial frameworks and the methods employed in such studies were still grounded in a view of humans as agents of disturbance. Similarly, though the idea of the ecology of the city as an integrated system is now being extended to focus on the entire urban area and its metabolism, that idea is still tied to the first conceptualization of systems theory and the emphasis on internal dynamics, which excludes relevant phenomena (O'Neill 2001).

A new approach to the study of urban ecosystems emerged as advances in both ecosystem science and urban ecology linked ecosystem processes to the system's heterogeneous structure. In urban ecology, this perspective is reflected in scholars' increasing interest in integrating ecology *in* cities with the ecology *of* cities (Forman 2014; Grimm et al. 2000; Pickett et al. 1997a,b). Conceptualizing the city as multiple equilibria and as a dynamic, highly unpredictable system is central to the current approach to urban ecology (Alberti et al. 2003).

During the past decade, scholars of urban ecology have started to expand their conceptual frameworks and methods of analysis to better represent socioecological interactions and explain emerging phenomena. To study urban ecosystems, we must integrate multiple boundaries— such as nested watersheds, land parcels, and natural patches—and analyze processes at multiple scales, ranging from metropolitan to local, regional, and global and including neighborhoods and households. We must also explicitly represent human agents and link the urban structure and human behaviors to ecosystem functions. Yet, although scholars do realize that socioecological interactions pose significant challenges to ecosystem science (Pickett, Cadenasso, and McGrath 2013), advances in urban ecology have not led to changes in ecological paradigms to fully integrate humans (Hixon, Pacala, and Sandin 2002; Reznick, Mateos, and Springer 2002; Robles and Desharnais 2002).

Eco-Evolutionary Dynamics on an Urbanizing Planet

A particularly significant challenge for ecology in the coming decades is to understand the role that humans play in eco-evolutionary dynamics. Humans are not only affecting ecosystem processes, they are altering the rules of nature's game by causing micro-evolutionary changes. Emerging evidence that evolutionary change can lead to ecological change over a relatively short timescale offers us the potential to significantly advance our understanding of the interplay between ecology and evolution in shaping environmental change (Schoener 2011). If rapid evolution does affect the functioning and stability of ecosystems, current rapid environmental change associated with urbanization and its evolutionary effects may have significant implications for ecological and human well-being over a relatively short period of time.

Such questions can be answered only by conducting major research in coming decades, but such investigations will require major revisions to the ways we study ecology and evolution. Eco-evolutionary feedbacks imply that populations modify their environment and that environmental changes influence the evolution of populations at comparable timescales (Post and Palkovacs 2009). Significant evidence shows that ecology affects selection on phenotypes (Endler 1986) and that organisms influence their environment through predation, nutrient cycling, and habitat modification (Odling-Smee, Laland, and Feldman 2003). Evidence of eco-evolutionary feedbacks over a relatively short timescale was postulated more than half a century ago (Pimentel 1961), but it has only recently been established (Schoener 2011).

I advance the hypothesis that as humans interact with niche construction and rapid evolutionary change through urbanization, they may alter the direction of evolution and the structures and functions of communities and ecosystems. I propose that integrating humans into eco-evolutionary science will provide important opportunities to understand the mechanisms of niche assembly on an urbanizing planet and the role that humans play in mediating eco-evolutionary feedbacks. Such feedbacks might arise at different scales in space and time and at different levels of biological organization (from genes to communities and ecosystems). Ecologists could test the effects of competition, predation, disease, and land-cover change on community organization; redefine Hutchinson's (1957) "realized niche" as an organism's hypervolume of occurrence in the presence of a gradient of human domination; and begin to explore the mechanisms by which humans affect eco-evolutionary feedback.

An Evolving Framework

A first step in advancing the field of urban ecology is to examine whether the assumptions of current frameworks limit our ability to explain ecological phenomena in urban ecosystems and to answer the key questions that they are specifically designed to address. The difficulty in such exploration, as O'Neill (2001) pointed out, is in remembering to consider the questions that existing frameworks might keep us from asking. Enough "anomalies" have accumulated in observations of urban ecosystem processes to challenge the fundamental assumptions we make in characterizing human-dominated ecosystems, and to indicate the need for a significant change in ecological research. The processes that contribute to urban development and ecology are extraordinarily complex, and many scholars have adopted diverse theoretical approaches to explain or predict them (Alberti et al. 2003; Grimm et al. 2000; Pickett et al. 2001; Pickett and Cadenasso 2002). Scholars of urban ecology have yet to develop a unified approach. Such a framework might diverge significantly from the initial conceptualizations that have characterized the history of urban ecology.

During the past few decades, various schools of urban ecology have developed different models to test hypotheses regarding the relationship between urbanization and ecosystem function. In my 2008 book (Alberti 2008), I reviewed the most influential perspectives that had characterized the initial development of a new urban ecology. Most dominant in the United States were teams of researchers in Phoenix, Baltimore, and Seattle (Collins et al. 2000; Grimm et al. 2000; Pickett et al. 2001; Alberti et al. 2003). These teams emphasized the unique interactions between patterns and processes and their human and ecological functions. The degrees to which these approaches integrated different disciplines vary with the specific composition of the teams constructing them. My colleagues and I proposed a conceptual framework that does not distinguish between human and ecological patterns and human and ecological processes (see figure 2.6). Our approach recognizes that patterns in urban landscapes are created by microscale interactions between human and ecological processes and that urban ecosystem functions are affected and maintained simultaneously by human and ecological patterns.

My research team and I saw urban ecosystems as complex, adaptive, dynamic systems (Alberti et al. 2003). Cities evolve as the outcome of myriad interactions among the individual choices and actions of many human agents (e.g., households, businesses, developers, and governments) and biophysical agents (e.g., local geomorphology, climate, and natural disturbance regimes). These choices produce different patterns of development, land use, and infrastructure density. They affect ecosystem processes both directly (in and near the city) and remotely through land conversion, use of resources, and generation of emissions and waste. These changes, in turn, affect human health and well-being (Alberti et al. 2003). Our approach explicitly recognized the "emergent" properties of the social-ecological systems that regulate system function. Yet the emphasis that all existing frameworks place on internal dynamics and feedbacks does not fully explain the relative stability of human-dominated systems that maintain ecosystem function as these systems evolve in response to continuous variability and change.

Regime shifts potentially associated with climate change—such as new hydrological regimes or the emergence of dead zones—pose serious challenges to the stability of urbanizing regions and increase their vulnerability (Miller et al. 2010). As humans dominate and alter Earth's ecosystems, they create new processes and mechanisms governing system dynamics. The likelihood of regime shifts may increase when humans reduce ecosystem resilience by modifying biogeochemical and hydrological cycles, reducing biodiversity, and changing natural disturbance regimes (Folke et al. 2004). A more complete framework of analysis explicitly represents multiple equilibria and regimes and system properties that allow urban ecosystems to adapt and evolve in response to change (figure 3.1).

The "complex systems" paradigm is a powerful approach for studying cities as emergent phenomena. Urban ecosystems are prototypical complex adaptive systems, which are open, nonlinear, and highly unpredictable (Folke et al. 2002; Gunderson and Holling 2002; Hartvigsen, Kinzig, and Peterson 1998; Levin 1998; Portugali 2000). Disturbance is a frequent intrinsic characteristic (R. E. Cook 2000). In such systems, patterns at higher levels emerge from the local dynamics of multiple agents interacting among themselves and with their environment (Nicolis and Prigogine 1989). Change has multiple causes, can follow multiple pathways, and is highly dependent on historical context; that is, it is path-dependent (P. M. Allen and Sanglier 1978, 1979; McDonnell and Pickett 1993). Agents are autonomous and adaptive, and they change their rules of action based upon new information. One of the least-understood aspects of urban development is the way that local interactions among

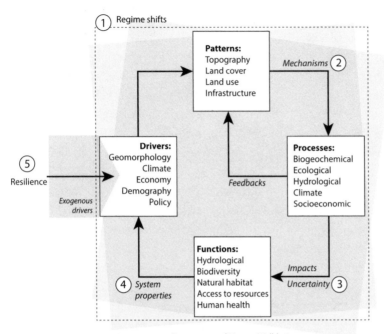

Ecosystem and Human Well-being

FIGURE 3.1   Resilience conceptual framework. In urban and urbanizing eco-systems, human and ecological functions and well-being are inextricably linked. Key human drivers of change are demographics, economics, technology, social organization, and political and governmental structures. Location preferences and consumption behaviors directly influence land use and demand on resources. Human drivers also interact with biophysical drivers (e.g., climate and topography), and the combination of these factors affects the heterogeneity of the landscape and its natural processes and disturbances. Modified from Alberti et al. 2003.

multiple agents can affect the global composition and dynamics of whole metropolitan regions.

Urban ecosystems can also be described as complex networks of human and natural agents connected by biogeophysical and socioeco-nomic processes (figures 3.2 and 3.3). Understanding the topology of interactions among components is an essential step in decoding urban landscapes and relationships between patterns and functions. Network theory is emerging as a promising approach to uncovering general rules of complex landscapes (Anderson, Frenken, and Hellervik 2006). To develop a theory of real networks, Albert and Barabási (2002) suggested

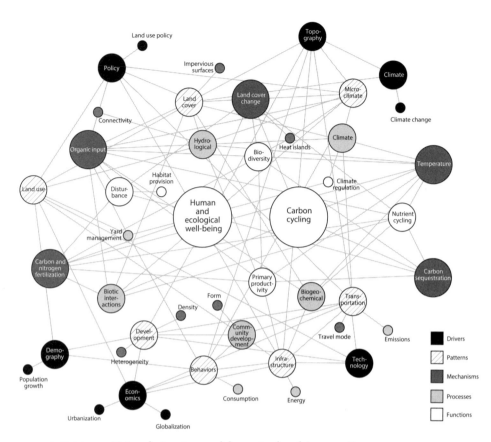

FIGURE 3.2 Network structure and dynamics: key drivers, patterns, processes, and mechanisms governing carbon function in urban ecosystems.

that we need to identify relevant parameters that, together with network size, yield a statistically complete characterization of the network. This implies developing a description of network dynamics (Albert 2006). To fully understand a complex landscape, we need to explore both the evolutionary path of the topology of socioeconomic networks and how that topology influences the ecosystem's dynamics and functioning.

The combination of real-time data acquisition and increased computing power is making it possible to investigate the topology and dynamics of networks containing millions of nodes and to explore questions that could not previously be addressed. But what has contributed enormously to the most recent advances in network theory is the crossing of disci-

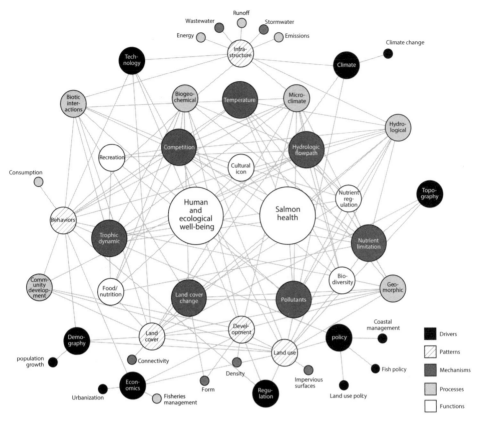

FIGURE 3.3   Network structure and dynamics: key drivers, patterns, processes, and mechanisms governing salmon function in urban ecosystems.

plinary boundaries and increasing access to data from many domains, which has helped scholars in diverse disciplines begin to identify generic properties of complex networks (Barabási 2005).

Rethinking the urban ecology framework has both scientific and practical implications. Considering some fundamental attributes of complex human and ecological adaptive systems—multiple interacting agents, emergent structures, decentralized control, and adaptive behavior—can help scholars study and manage urban sprawl as an integrated human-ecological phenomenon. The emerging urban landscape structure can be described as a cumulative and aggregate order that results from many locally made decisions involving an array of intelligent and adaptive agents. Complex metropolitan systems cannot be managed by a single set of top-down government policies (Innes and Booher 1999); instead, they require that multiple independent players coordinate their activities under locally diverse biophysical conditions and constraints, constantly adjusting their behaviors to maintain an optimal balance between human and ecological functions.

Anomalies of Urban Ecosystems

Urbanizing regions present challenges to ecosystem ecology, but scholars in the field can gain important insights from studying these areas. Searching for explanations of the unique ecological phenomena that we observe in urban ecosystems, scholars of urban ecology still apply assumptions heavily dominated by the old ecosystem paradigm. We see evidence of the persistent old paradigm in the bias toward an ecosystem approach that separates ecological and biophysical drivers from human ones and emphasizes human impact on ecological processes, disregarding the fact that humans participate fully in the ecosystem functioning of such settings. Thus ecosystem functions in urban ecosystems, and the associated phenomena of productivity, homogenization, and stability, are typically explained, at best, as unintended consequences of human activity. In such explanatory models, humans are still external to the ecosystem and participate in its dynamics only as a driving force.

An alternative explanation for ecological phenomena that have been associated with urbanization, such as biotic homogenization (Groffman et al. 2014) or the urban stream syndrome (Walsh, Fletcher, and Ladson 2005), is the fact that urban and suburban areas are "human habitats" and are designed and managed to best support human functions. Urban ecology lacks a theory and science of human habitats comparable to those of other species' habitats. Often we study cities as ecological systems, disregarding the fundamental fact that they are built for humans and their well-being. Cities' socioeconomic functions are what drive microclimate change and the composition of species diversity, phenomena that lead to more homogeneous biophysical environments (McKinney 2006). As urbanization increases, biotic homogenization probably does so, too. As cities expand, urban regions are maintained in a state of disequilibrium from the local natural environment, so that habitats across urban sites are more similar to one another than to their respective adjacent natural environments (Groffman et al. 2013).

For a new paradigm of urban ecology to emerge, scientists have to develop a theoretical framework that explains the anomalies emerging from observations in urban ecosystems, and they have to articulate testable hypotheses of how urban ecosystems and urbanization patterns affect relationships among biodiversity, ecosystem processes, and resilience. Understanding the relationship between species diversity and ecosystem function and stability on an urbanizing planet is critical. Pickett et al. (2008) have illustrated findings from the Baltimore Ecosystem Study (BES) that question assumptions often made by ecological scholars studying urban areas. I suggest that such observations might not be reconciled with the current ecological paradigm, and I support this suggestion by highlighting a few examples of observations in urban ecological studies that cannot fully be explained by existing theories that focus on biodiversity, hydrology, biogeochemistry, and evolution. Scholars of urban ecology need to develop empirically based novel theories to fully integrate humans into ecosystem science.

Biodiversity

Urban development affects biodiversity both directly, by altering land cover and introducing nonnative species, and indirectly, by changing ecosystem and biogeochemical processes (Alberti 2008). Human activities reduce the number of species and selectively determine the diversity of phenotypic traits. Since individual species may control community- and ecosystem-level processes (Lawton 1994; Paine 1984; Power et al. 1996), diversity may have a strong effect on ecosystem function; changing diversity affects the probability that these influential species will occur among potential colonists (Cardinale, Nelson, and Palmer 2000; Tilman 1999). Humans can influence ecosystem processes both by altering the dominance of species with particular traits and by facilitating or impeding complementarity among species with different traits. Findings from current urban ecological studies at the two urban Long Term Ecological Research (LTER) sites in Baltimore and Phoenix, as well as from Seattle, provide important evidence that allows scholars to develop hypotheses on the mechanisms that drive urban biodiversity. These include habitat productivity, species interactions, trophic dynamics, heterogeneity, disturbance, and evolution.

In urban ecosystems, the numbers of synanthropic species (animals and plants that benefit from humans) are directly related to the degree of urbanization of a given area; in the upper range of the urban gradient, synanthropic species outcompete native ones. People directly affect species selection and create new vectors of seed dispersal. Both directly and indirectly, urbanization modifies natural disturbances (e.g., fire, pest outbreaks, etc.) and gives rise to new ones (e.g., chronic stresses, unnatural physical structures, synanthropic species, and degrees of connectivity), leading to the emergence of novel habitats. Urban landscapes are also different in their patch dynamics (Pickett and Rogers 1997; Wu and Loucks 1995) and heterogeneity (Pickett and Rogers 1997).

Ecological scholars have started to investigate how spatial heterogeneity in urban regions influences the flow of energy, materials, species, and information across the urban landscape. Machlis, Force, and Burch (1997) described the urban landscape as a complex mosaic of biological and physical patches within a matrix of infrastructure and social organization. Spatial heterogeneity in an urban ecosystem is generated by both biophysical and human processes (Pickett, Cadenasso, and Jones 2000). High numbers of small patches with differing environmental conditions could suggest a relatively high species diversity in urbanizing regions, since these areas can simultaneously support a variety of native and introduced plant species (Gilbert 1989; Kowarik 2011; Pyšek 1993; Sukopp, Blume, and Kunick 1979). Yet empirical studies suggest that habitat changes associated with urban land uses act as a filter for urban species composition, with clear winners and losers.

Based on these findings, scholars have advanced the hypothesis that urbanization drives the homogenization of ecological structures and functions. This process has significant implications for ecosystem carbon and nitrogen dynamics at the global scale. The phenomenon of homogenization is complex, however, as it reflects a nonlinear environmental response to disturbance. Several hypotheses emerging in the literature have yet to be tested. McKinney (2006) observed that while synanthropic species adapted to intensely modified built habitats are "global homogenizers," species diversity and abundance in such habitats are often as great as, or greater than, in the surrounding landscape because of native and early successional species that occupy suburban habitats.

Such complexity, both physical and biological, is inextricably linked to the complex social dynamics between urban and suburban communities and across biophysically dissimilar regions. Groffman et al. (2013) hypothesized that neighborhoods whose residents have similar lifestyle characteristics (e.g., age, socioeconomic status, life stage, and ethnicity) and social preferences (e.g., values and interests) across biophysically dissimilar regions have more similar landscaping practices (e.g., watering and fertilization) than do socially dissimilar neighborhoods within the same city. Yet when researchers examine biophysical outcomes at different scales, their evidence suggests that specific behaviors around land management practices (e.g., lawn care) are differentiated (Polsky et al. 2014).

As we explore the relationship between biodiversity and ecosystem function in human-dominated ecosystems, it becomes evident that humans influence the pattern of biodiversity as well as the ecosystem functions that biodiversity provides. Emerging evidence does converge on key mechanisms governing the way that biodiversity responds to human drivers, but no one has yet developed a unified theory. Current ecological hypotheses can only partly explain the patterns and processes of urban biodiversity. Mutual interactions and feedback occur among biodiversity changes, ecosystem functions, and abiotic factors. Humans influence all these factors by strengthening or loosening some of these interactions and feedback loops and by creating unprecedented interactions between biodiversity and ecosystem function. Thus the relationships that govern urban ecosystem diversity and stability cannot be understood outside the context of these complex human-influenced interactions (Ives and Carpenter 2007).

When studying the impact of humans on biodiversity and ecosystem function, it is also critical to acknowledge that processes influencing diversity operate at different scales of space and time (MacArthur 1969; Tilman and Pacala 1993). Humans influence these dynamic interactions across a wide range of spatial and temporal scales through urbanization and land-cover change, which potentially alter the dynamic of community assembly. Furthermore, recent studies have pointed out that important feedback mechanisms might link ecosystem function and biodiversity across such scales (Hughes et al. 2007; Loreau 2010).

Niche Construction. Integrating humans into the study of biodiversity may also suggest a way to reconcile key concepts in both niche theory and the theory of dispersal-based community assembly. Early on, my team and I (Alberti et al. 2003) noted opportunities to advance ecological theory by integrating humans into the study of ecosystems. Hutchinson (1957) transformed and solidified the concept of niche, changing it from a mere description of an organism's functional place in nature (Elton 1927) to a mathematically rigorous n-dimensional hypervolume that could be treated analytically. Hutchinson's "realized niche" included only those places where an organism's physiological tolerances were not exceeded (its "fundamental" niche) and where its occurrence was not preempted by competitors. In addition to competition, other forces may be important to the organization of communities, such as predation, resource variability, and productivity at large scales (e.g., climate and disturbance regimes).

By integrating humans into the study of processes controlling biological diversity, ecological scholars may also be able to resolve important puzzles in island biogeography and explain empirical results regarding the balance between colonization and extinction in human-dominated ecosystems (Marzluff 2005). As humans urbanize, they cause the emergence of new selective forces, including new habitats, disturbance regimes, predators, competitors, and diseases that may drive native species to extinction (Kühn, Brandl, and Klotz 2004; Olden and Poff 2004; Sax and Gaines 2003).

Species-Area Relationships. Within the debate on biodiversity, another area of contention focuses on species-area relationships: the idea that the number of species increases with the size of the sampling area, such that larger areas contain more species (Arrhenius 1921). This relationship is important for both ecosystem science and management. It is the basis for adequately sampling the species in a particular community, characterizing community structure, and estimating species richness (Connor and McCoy 1979). For conservation biology, it also provides guiding principles to define the optimal size of reserves (He and Legendre 1996).

Urbanizing landscapes provide unique opportunities to expand our understanding of the species-area relationship, to apply the knowledge to improve design, and to better manage urban regions.

Disturbance. In ecology, a disturbance is any relatively discrete event in space and time that disrupts ecosystem, community, or population structure (White and Pickett 1985). Disturbances influence the coexistence of species and the maintenance of biodiversity. The Intermediate Disturbance Hypothesis predicts that an intermediate frequency or intensity of disturbance maximizes diversity. At a high frequency of disturbance, species diversity is low because only "weedy" species that quickly colonize and reach maturity are able to survive. At the other extreme—low disturbance frequency—species diversity is expected to be low because competitively dominant species exclude those that are less competitive. Only at intermediate levels of disturbance does a mix of colonizers and competitors coexist (Connell 1978; Petraitis, Latham, and Niesenbaum 1989; Sousa 2001).

At the same time, diversity can mediate the severity of disturbances, leading to variation in the frequency, intensity, or duration of actual biomass loss (i.e., realized disturbance) across communities with differing initial levels of diversity (Hughes et al. 2012). Researchers have found significant evidence of reciprocal effects between disturbance and diversity and between diversity and community stability (Allison 2004; Cardinale and Palmer 2002; Mulder, Uliassi, and Doak 2001; Tilman 1996). Diversity influences the response to disturbance by increasing the probability that a given species can compensate for negative responses of other species (the insurance hypothesis; e.g., Tilman 1996). Furthermore, biodiversity provides spatial insurance for ecosystem functions by facilitating spatial exchanges among the local systems that compose heterogeneous landscapes.

Urbanization modifies existing disturbance regimes both directly (e.g., fire and flood management) and indirectly (e.g., microclimate alteration). To create a desirable human habitat, humans suppress disturbances such as fire or floods; the result is reduced heterogeneity in natural features (M. G. Turner, Carpenter, and Gustafson 1998). Furthermore, changes in microclimates, hydrological patterns, geomorphology, soil conditions, and habitat indirectly modify natural disturbance regimes.

In addition, by changing biodiversity, urbanization interacts indirectly with the spatial insurance effects of species diversity that (as I described above) are hypothesized to be highest at the intermediate dispersal rates that maximize local diversity. These results have profound implications for understanding the effects of urban landscape heterogeneity and for pursuing conservation and management efforts.

Hydrology

Urbanization has significant effects on the hydrologic cycle, as it results in construction of infrastructure intended to transfer and control water flow. Extracting water to meet the needs of urban residents and their activities affects flow regimes in urbanizing watersheds (Churkina, Brown, and Keoleian 2010; Pataki et al. 2007; Pickett et al. 2011). Urbanization implies an increase in impervious land area, which affects both geomorphological and hydrological processes, thus causing changes in water and sediment fluxes (Booth and Jackson 1997; Wolman 1967; Booth and Fischenich 2015). In cities, water moves over primarily impermeable surfaces, traveling rapidly and collecting a variety of human-made organic and inorganic pollutants before it reaches a treatment plant or body of water. A complex network of culverts, gutters, drains, pipes, sewers, and channels extracts, redistributes, and moves water between the sites of the natural hydrological processes and those of the artificial hydrological systems that support human settlements and activities. Compared with the drainage basins of forested areas, urban basins with 10 to 20 percent total impervious surface may have twice as much surface runoff and significantly shorter lag times between precipitation input and discharge. They also generate higher flood peak discharges during storms (Paul and Meyer 2001). Furthermore, due to stream incision patterns, water tables tend to be lower in urbanized areas than in forested ones (Groffman and Crawford 2003).

Another significant influence on water flow in urban watersheds is channelization: humans straighten, deepen, and/or widen stream channels to prevent flooding and facilitate urban drainage (Arnold, Boison, and Patton 1982; Booth 1990; Klein 1979). Channelization modifies flow regimes over time. Urban development affects geomorphology at various stages (Booth 1990; Booth and Jackson 1997; Dunne and Leopold 1978). For example, bridge construction can directly alter stream channels while also restricting stream flow, reducing flow velocities upstream, and increasing sedimentation. In addition, urban development disturbs soil, increases movement of sediments, and removes vegetation from stream banks (Jacobson, Femmer, and McKenney 2001). Such direct and indirect changes to stream habitats can have dramatic effects on aquatic organisms (Walsh, Fletcher, and Ladson 2005) and affect the distribution of stream processes, including leaf decomposition, in freshwater communities (Webster and Benfield 1986): when vegetation is lost, stream temperatures can rise, in turn altering many elements of in-stream ecology.

Although most urban streams display many characteristics of "urban stream syndrome," they vary a great deal across cities and urbanizing regions (Walsh, Fletcher, and Ladson 2005). Emerging hypotheses to explain such differences range from hydrological variability and biophysical interactions to land use legacies and human behaviors. The extent to which development patterns and infrastructure play roles in explaining this variability has not been fully investigated.

Biogeochemistry

Recent studies of biogeochemistry in urbanizing regions provide evidence of the complex mechanisms by which urban activities affect local and global biogeochemical cycles (Churkina, Brown, and Keoleian 2010; Pataki et al. 2007; Pickett et al. 2011). As humans have begun to dominate Earth's ecosystems, we have mobilized large amounts of nitrogen, phosphorus, and metals from the Earth's crust and atmosphere. These alterations to biogeochemical cycles feed back to affect the climate and the redistribution of nutrient and mineral resources worldwide, with significant consequences for ecosystem functioning (Grimm et al. 2013).

A few empirical studies have started to quantify relationships between urbanization and nutrient cycling (Grimmond 2006; Gurney et al. 2009; Hutyra, Yoon, and Alberti 2011; Nowak and Crane 2002); however, available data are too limited to formally test hypotheses about the underlying processes and mechanisms linking patterns of urban development to the global cycles of carbon (C), nitrogen (N), and phosphorus (P). The magnitudes of, and mechanisms governing, nutrient fluxes in urbanizing regions remain highly uncertain because of complex interactions between human and ecological systems and because of the effects that diverse human behaviors and built structures have on biophysical processes and dynamics (Pataki et al. 2006). In addition, we do not know how alternative patterns of urbanization may influence human behaviors, biogeochemistry, and ecological response.

In the absence of empirical data, current models of the C cycle assume that whereas urban C stocks and fluxes may vary in magnitude, they are controlled by generally understood mechanisms of global C dynamics (Canadell et al. 2003). However, initial empirical analyses across gradients of urbanization suggest that interactions between human activities and biophysical processes may create new processes and mechanisms governing ecosystem dynamics (Kaye et al. 2006). C exchanges may be significantly influenced by competing positive feedbacks (e.g., heat islands or N fertilization) and negative ones (e.g., declines in forest cover or removal of leaf litter and its associated nutrients), which create distinctive patterns and processes across a gradient of urbanization. Scholars of urban ecology have posited a series of plausible hypotheses about a biogeochemistry unique to urban areas, which may vary across regions and over time (ibid.).

The urban C cycle is evolving, complex, and not fully understood. Lucy Hutyra and I synthesized findings from observational studies in U.S. metropolitan areas by focusing on five key mechanisms that affect change in C stocks and fluxes along a gradient of urbanization: land-cover change, emissions, organic inputs, temperature, and N fertilization (Alberti and Hutyra 2013). Variations in the C budget across an urban gradient result from the influence of the gradient on several mechanisms that govern C uptake, release, and storage. As described in detail in chapter 5, Hutyra and I hypothesize that these five mechanisms produce nonlinear variations in C stocks and fluxes across the urban gradient. The amount of C in vegetative biomass (and soils) is expected to increase with decreasing intensity of development, with small peaks in the older suburbs and exurbs where larger lots have had time to accumulate biomass after their initial clearing. Fluxes (per-unit mass) might be expected to decrease with decreasing temperatures and decreased N and CO_2 fertilization, but ultimately they will be highest in the least dense areas because of the large amount of photosynthetically active vegetation in forests.

The hypothesized relationships between urbanization and the C cycle highlight the complexity of the relationships that need to be investigated in order to fully characterize the functioning of urban ecosystems (Hutyra et al. 2014). We also expect that these relationships vary across biomes and regional socioeconomic settings. Cross-comparative studies are required to understand how urban patterns affect these relationships. Urban regions are rapidly expanding, and, in many respects, urban expansion is critical to the advancement of viable scenarios for net emissions reduction. Studying how patterns of development mediate their impact could provide important information for urban planning and management (see chapter 5).

Eco-Evolutionary Feedbacks

The significant decrease in biodiversity in cities is only the most apparent of several more subtle changes that are associated with urbanization and that have the potential to affect genetic diversity (Marzluff 2012; Partecke 2014; Pickett et al. 2011). Urbanization alters natural habitats, leading to new selection pressures and phenotypic plasticity. The new selection regimes have significant consequences for microevolutionary changes. Habitat modification and fragmentation may lead to genetic differentiation. New predators and competitors affect species interactions. At the same time, extreme turnover in biological communities might prevent the genetic differentiation of urban populations and impede evolutionary responses to the novel selective forces associated with urbanization (Shochat et al. 2006).

There is increasing evidence that human-driven microevolutionary change may affect the population dynamics of natural organisms through survival or reproductive success, leaving a genetic signature with a cascade of effects on community dynamics and ecosystem functions (Hendry, Farrugia, and Kinnison 2008; Palkovacs et al. 2012; J. N. Thompson 1998). Genetic signatures have been observed in the population dynamics of several organisms, including birds, fish, arthropods, rodents, land plants, and algae (Hendry et al. 2008; Palkovacs and Hendry 2010).

The challenge for urban ecology in the next decades is to understand the role humans play in eco-evolutionary dynamics (Alberti 2015). If rapid evolutionary change affects ecosystem functioning and stability, the current rapid environmental change driven by urbanization may have significant implications for ecological and human well-being on a relatively short time scale. Humans challenge the cultural and genetic makeup of species on the planet. At the same time, completely novel interactions between human and ecological processes may produce opportunities for innovation. In chapter 6, I identify emerging hypotheses on how urbanization drives eco-evolutionary dynamics. Understanding the mechanisms by which cities mediate evolutionary feedback will provide insights to guide the maintenance of ecosystem functions over the long term.

Elements of an Emerging Paradigm

This chapter has highlighted empirical observations that challenge urban ecologists to expand ecological theories and revise key concepts by including the perspective of humans, the dominant species in urban ecosystems. A co-evolutionary paradigm may be more appropriate to represent urban ecosystem dynamics. In urban ecosystems, humans are coproducers of ecosystem function. As full participants, humans have functions that are uniquely associated with their traits and are likely to produce urban ecosystem functions that no other species can carry out. Humans can offer no substitute for the ecosystem functions provided by plant species (such as primary production and nutrient or hydrological cycling), but they do support these functions indirectly, in human-dominated systems, as they consume resources and modify the physical structure and chemical composition of the environment. In urban ecosystems, humans also support new ecosystem functions that promote human health and well-being (e.g., food and drinking water, moderate climate, and aesthetic and recreational values).

The concept of ecosystem services has emerged over the past thirty years and identifies benefits that natural ecosystems provide to human society (Daily et al. 1997). The concept evolved with an explicit emphasis on the natural end of the spectrum, focusing on services flowing from natural ecosystems to support human function—thus avoiding the distinction between human and natural ecosystems. Yet in its original conceptualization, humans are not considered to be service providers, but only end users of such services. In contrast, I suggest that in human-dominated ecosystems, human functions are essential to providing and maintaining ecosystem functions for both nature and society.

Unless we expand the definition of ecosystem function in urban ecosystems, the role that humans play will be only partially represented, and will be misrepresented overall, in the current view of ecosystem services. In ecology, humans will continue to be seen and studied simply as destructive agents of system function, under the assumption that the system "normally" operates as a "pristine" natural state. An example is arrested or modified successions, in which the natural process of ecological succession is halted by human intervention. Following the removal of trees by deforestation, the subsequent management of land and established disturbance regime maintain the ecosystem in a state of disequilibrium, preventing the return of the forest. But where is the place of humans in such a system?

The patterns, processes, and dynamics governing ecosystem functions in urban ecosystems are poorly understood, and these ecosystems' management is grounded in the assumption that they can operate and function like ecosystems of the same type under undisturbed conditions. In contrast, Krebs (1988) suggested that ecosystems can exist in more than one stable (i.e., resilient and healthy) configuration, which poses the question of whether there is a unique domain of health for a given ecosystem type. If more than one domain of ecosystem functionality exists within the constraints set by intrinsic environmental conditions, this would imply that each possible stable configuration can deliver a different set of ecosystem functions. Deciding which set to preserve becomes a question of societal preference (Tett et al. 2013).

In chapter 2, I suggested that coupled human-natural systems are hybrid ecosystems: they fluctuate between two different basins of attractions or alternative states, characterized by historical versus novel feedback mechanisms. Under such unstable conditions, ecosystem functions evolve from their state in an undisturbed natural system to their state in a human-dominated one. Trade-offs may emerge between the processes that support such functions. Managing ecosystems under assumptions of pristine conditions may drive unintentional systems to collapse. For example, while urban forests can mitigate storm water flows, remove pollutants, and cool the urban environment, they also have costs, such as high water requirements. Urban forests are complex and costly for cities to create, maintain, and protect, and vegetation and associated benefits are often unevenly distributed. This implies potentially complex trade-offs as water resources become scarcer.

The Dynamics of Multiple Equilibria Systems

Holling (1996) pointed out four key characteristics of ecological systems:

1. Ecological systems are complex, dynamic, and open and have multiple equilibria. These characteristics of urban ecosystems have important implications for defining their multiple states and for evaluating the effects of urban patterns on the systems' abilities to maintain human and ecological functions over the long term. Feedback mechanisms can amplify or regulate a given effect.

2. Change is neither continuous and gradual nor consistently chaotic. Rather, it is episodic, with periods of slow accumulation punctuated by sudden reorganization. These events can shape trajectories far into the future. Critical processes function at very different rates, but they cluster around dominant frequencies. Episodic behaviors are caused by interactions between variables that have immediate versus delayed effects.

3. Spatial attributes are neither uniform nor scale-invariant. Instead, they are patchy and discontinuous. Therefore, scaling up is not a simple process of aggregation, as nonlinear processes determine how the shift occurs from one scale to another. This observation has implications for understanding the effects of spatial interactions between human and ecological systems at multiple scales.

4. Ecosystems are moving targets. Knowledge is incomplete, and surprise is inevitable. This observation has implications for the type of strategies we adopt. Instead of fixed policies, perhaps we should consider flexible mechanisms and governing institutions that can learn effectively and deal with change.

The Discontinuity Hypothesis

Ecological structures and processes occur at specific spatio-temporal scales, and interactions operating across multiple scales mediate scale-specific (e.g., individual, community, local, or regional) responses to disturbance (Nash et al. 2013). The discontinuity hypothesis provides an important avenue for understanding both cross-scale interactions within and across human and natural systems—by linking ecosystem function to habitat structure—and their inherent scale within ecosystems.

Discontinuities within ecosystems can help to explain relationships between habitat structure and attributes of associated communities as well as emergent phenomena such as resilience (Brown 1995; MacArthur and Wilson 1963; Milne et al. 1992; Ritchie 1998). Thus examining discontinuities in habitat structure offers a fertile avenue for understanding coupled human-natural processes and multiscale interactions by allowing identification of scale-specific relationships among ecosystem drivers and the human and natural processes that characterize urbanizing landscapes.

Studying the boreal forest in Canada, Holling (1992) observed a correlation between breaks in distributions of animal body masses and discontinuities in forest structures and processes. He proposed that where ecosystem patterns are persistent over ecological timescales, biological processes unrelated to the original structuring processes become entrained by, and adapted to, those patterns across scales. For example, attributes of animals, including their life histories, behaviors, and morphologies, may adapt to a discontinuous landscape pattern, as this pattern reflects opportunities for shelter, food, and resources (Fauchald and Tveraa 2006). However, these opportunities are mediated by the scales at which individuals interact with the landscape and exploit resources (Haskell, Ritchie, and Olff 2002; Holling 1992), and the scales of these interactions are positively correlated with body size (Peters 1983).

Various studies looked at a number of species and community attributes, such as species' abundance and biomass (Angeler et al. 2011), richness (Warwick, Dashfield, and Somerfield 2006), and range size (Restrepo and Arango 2008). Studies across spatial and temporal scales show similarly discontinuous distributions in social systems. Despite increasing evidence that temporal and spatial discontinuities are fundamental elements of system structure and function, the specific mechanisms driving the link need to be explored explicitly (Robson et al. 2005).

Resilience

The complexity of coupled human-natural systems results from the interactions of many qualitatively distinct entities across different levels of the biological and social hierarchy in processes that cross multiple scales in space and time (Holling 1992; Levin 1992, 1999; Peterson, Allen, and Holling 1998). Increasing evidence from natural and social systems suggests that their structure and dynamics are influenced by biotic, abiotic, and human processes operating at specific temporal and spatial scales (Holling 1992; O'Neill et al. 1986). At each level, the difference in speed and size may introduce discontinuities in the distribution and patterns of key ecosystem attributes (T. F. H. Allen and Starr 1982; O'Neill et al. 1986). This difference has important implications for ecological structure and the roles of specific processes operating at a given scale (Nash et al. 2014).

Scholars have recently hypothesized that these structuring processes produce discontinuities in ecosystems; they have provided a framework to better identify which structuring processes are relevant and to assess system resilience (C. R. Allen, Gunderson, and Johnson 2005; Holling 1992). New research suggests that many aspects of cities may be governed by universal properties related to their size and driven by human interactions (Bettencourt 2013). The relationship of these "scale laws" to the resilience of urbanizing systems has yet to be investigated deeply, but it offers significant promise for guiding urban development in response to environmental variability and change. Kerkhoff and Enquist (2007) provided some initial hypotheses by focusing on observations of settlement size and fertility rates. They suggested that deviations from scaling relationships in ecological and social systems may be the signature of specific structuring processes or reorganizations of the system. Such patterns provide the basis for developing principles that we can apply to better understand—and better manage—coupled human-natural systems.

Understanding the co-evolution of human and natural systems is key to building a resilient society and transforming our habitat. To understand how evolution constructs the mechanisms of life on a human-dominated planet, we must understand how human and natural systems self-organize and how their interplay influences evolutionary change. Reframing urban ecology implies not simply expanding the current framework and approaches, but also acknowledging the plurality of sciences that form the essential structure of the field. Future policies and management practices will succeed or fail based on their ability to acknowledge and address the complexities and uncertainties of these systems.

4

Emergent Properties in Coupled Human-Natural Systems

URBAN ECOSYSTEMS ARE QUALITATIVELY DISTINCT FROM OTHER environments. In such systems, change and evolution are governed by complex interactions among ecological and social drivers. In this chapter, I investigate the emergent properties that characterize urban ecosystems by focusing on patterns (e.g., sprawl), processes (e.g., hydrology), and functions (e.g., flood regulation). I develop an analytic approach to examine complex interactions between slow and fast variables that control critical transitions, regime shifts, resilience, and innovation. I articulate formal hypotheses regarding relationships between emergent properties of hybrid ecosystems and their abilities to adapt and innovate. Uncertain future interactions between social and ecological dynamics call for a paradigm shift in urban design and planning.

Emergent Properties of Urban Ecosystems

Urban ecosystems are highly complex. They are hybrid, open, nonlinear, unpredictable, and are characterized by multiple equilibria (Folke et al. 2002; Gunderson and Holling 2002; Hartvigsen, Kinzig, and Peterson 1998; Levin 1998; Portugali 2000). Hybrid ecosystems are characterized by complex interactions, emergent properties, and dissipative thermodynamics (Nicolis and Prigogine 1977). At higher levels, patterns emerge from the local dynamics of multiple agents interacting among themselves and with their environment. Sprawl is an example of an emergent pattern of dynamic interactions among human preferences for residential location, individual mobility patterns, transportation infrastructure, and real estate markets, and between these factors and the regional climate, hydrology, and topography (Alberti 2008; Torrens and Alberti 2000). Households, which are themselves complex entities, compete simultaneously in the job and real estate markets when people decide where to live. Furthermore, people have preferences and evaluate trade-offs that are highly dependent on their individual characteristics (e.g., income, household size, and children) and the socioeconomic (e.g., quality of public services) and environmental (e.g., environmental amenities) attributes of potential alternative housing locations.

Developers and local governments make decisions about the development of land and infrastructure. These decisions are strongly influenced by consumer preferences and housing demand and are also shaped by biophysical (e.g., topography), economic (e.g., resources), and institutional (e.g., property rights) constraints. Metropolitan patterns eventually emerge from local interactions among various agents and their decisions; in turn, these patterns affect processes that support urban ecosystem functions (e.g., mobility, air quality, natural habitat, safety, etc.). Resulting changes in urban ecosystem functions feed back into choices about households' locations.

In such systems, change and evolution emerge as various interacting agents engage in simple behaviors and as they respond to external factors. Uncertainty about future conditions is important, as any departure from past trends can affect how a system evolves. Uncertainty and the likelihood of surprise are controlled by complex interactions among ecological and social drivers and their unpredictable dynamics (Alberti 2008). Change has multiple causes, can follow multiple pathways, and is highly dependent on historical context; that is, it is path dependent (P. M. Allen and Sanglier 1978, 1979; McDonnell and Pickett 1993). Agents are autonomous and adaptive, and they change their rules of action based on new information. In urban ecosystems, feedback mechanisms that operate between ecological and human processes can amplify or dampen changes, and thus they regulate the systems' responses to external pressures. For example, in delta regions, land-cover changes and rapid loss of tidal marshes, coupled with hydrological and ecological changes associated with the development of hard flood-control structures (e.g., dikes, dams, levees, groins, seawalls, and stormwater management measures), make systems more vulnerable to extreme climate events and prompt increased demand for more flood-control infrastructure (box 4.1).

Urban ecosystems are also highly interdependent social and ecological networks (i.e., networks of networks). In cities, built and virtual infrastructures act as networks of systems that produce and distribute a continuous flow of essential services. As in biological systems, where genetic regulatory and protein interaction networks are interdependent, so too in human systems do infrastructure networks depend on one another to function. For example, the power grid is highly dependent on communication networks for its control, but the communication network also depends on the grid for power (Parandehgheibi and Modiano 2013). Coupled human-natural systems are composed of overlapping and interdependent networks. As for ecological systems, the heterogeneity and connectivity of system components play a critical role in maintaining functional systems. Increasing connectivity among people, business enterprises, and governmental and nongovernmental organizations around the globe enables communication, efficiency, and innovation. At the same time, this increasing interdependence may increase vulnerability, instabilities, and the potential for effect cascades.

These systems evolve through feedback loops, nonlinear dynamics, and self-organization (Nicolis and Prigogine 1977). A fundamental characteristic of urban systems is their dissipative thermodynamics. Cities maintain their internal order by using matter and high-quality (low-entropy) energy from external sources, which allows the system to maintain itself and evolve to a higher degree of complexity and order (Prigogine 1978). A key property of self-organized systems is criticality— a threshold state between stability and instability (Bak, Tang, and Wiesenfeld 1987). The concept of criticality is particularly relevant when thinking about system evolution, environmental change, and adaptation of urban ecosystems. Cities are dynamic systems operating at the border of a phase transition between two behavioral regimes. According to Kauffman (1993), criticality facilitates the emergence of complex aggregated behaviors that reach an optimal balance between stability and adaptability. Critical systems maximize their ability to use information about their past and innovate to respond to future conditions. In hybrid ecosystems, novelty emerges from interactions between human and natural functions.

Box 4.1. Infrastructure as a Regulator of Urban Resilience

Interactions between human and natural drivers may generate regime shifts, such as flooding, loss of estuary biodiversity, and algal blooms. Regime shifts are shifts in dominant feedback. In urbanizing coastal regions, they emerge from complex interactions among natural processes and human interventions, which modify feedback mechanisms that maintain system functions and that prevent the system from shifting to alternate states. Historically, river deltas would shift location and expand and contract—depending on changes in precipitation and erosion—within a stability domain. Slow changes

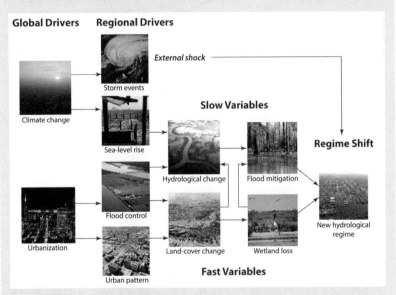

FIGURE B4.1 Slow and fast variables. Urban patterns have complex impacts on variables that change slowly in response to long-term processes (e.g., the capacity of soil to retain water and mitigate floods in coastal regions) and that constrain the response of fast variables (e.g., land cover change and wetland loss), generating a shift in the hydrological regime. Photographs: climate change: Michel; urbanization: Brandon Doran; storm events: NASA Goddard Space Flight Program; sea level rise: Florida Sea Grant; flood control: Tobin; urban pattern: Sam Korson; hydrological change: Feral Arts; land cover change: USACE; flood mitigation: Yortw; wetland loss: Coastal Wetlands Planning, Protection, and Restoration Act; new hydrological regime: aerial photo of Katrina, National Oceanic and Atmospheric Administration.

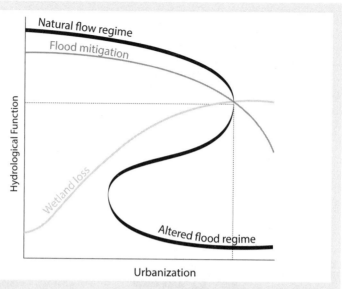

FIGURE B4.2   Regime shift in hydrological systems. Marina Alberti.

in internal feedback caused by hardening of banks and levees may change the domains of attraction from a resilient condition to a less resilient one, thus changing the thresholds that push the system into a new regime.

Urban patterns have complex impacts on what resilience scientists define as *slow* and *fast variables* (figure B4.1). Slow variables (such as the capacity of soil to retain water and mitigate flooding in coastal regions) constrain the response of fast variables (such as land cover change and wetland loss), leading to potential shifts in the hydrological regime. The fast growth and rapid land transformation caused by urbanization, together with sea-level rise and the increase in size and frequency of flooding events associated with climate change, may be a tipping point in the emergence of a new hydrological regime in urbanizing deltas across the world (figure B4.2).

When multiple phenomena with uncertain trajectories (e.g., storm surges and power outages) simultaneously affect multiple cities' functions, the element of surprise can be enormous. Resources and activities that are normally taken for granted, such as mobility or an energy supply, may suddenly become unavailable, causing ripple effects on human safety and well-being. In a comprehensive report developed in 2011 for the New York State Energy Research and Development Authority (NYSERDA), a team of climate scientists predicted that given the expected increase in extreme events such as hurricanes, subway tunnels in New York could flood, as they in fact did during Hurricane Sandy (Rosenzweig et al. 2011). Far less predictable, however, were the many interactions of rail and road closures and potential consequences for the thousands of city dwellers who had to shift their mobility patterns, especially while power outages and other system failures simultaneously limited access to food supplies and drinking water.

Anticipating critical transitions, regime shifts, and failures in complex systems is one of the most valuable contributions that resilience science can provide decision-makers. Emerging studies of complex, interrelated networks reveal fundamental architectural features that differentiate systems that have tipping points—which can be triggered by extreme events—from those that can adapt. In addition to diversity and modularity, other properties hypothesized to enhance adaptive capacity are cross-scale interaction, early warning, and self-organization. Early warning becomes especially valuable when dealing with irreversible state changes, as the ability to delay or prevent these changes is far more desirable than trying to force a return to the previous state. Emerging research on early warning points to signals within complex systems that can help detect looming thresholds and regime shifts.

Four properties, known as the 4 Rs, have been proposed by the Multidisciplinary Center for Earthquake Engineering Research (MCEER) to specifically assess the resilience of an infrastructure system:

1. *Robustness* is the ability of systems or system components to withstand potentially disastrous forces without significant degradation or loss of performance.
2. *Redundancy* is the extent to which systems or system components can be substituted for one another if significant degradation occurs.
3. *Resourcefulness* is the ability of a management system to diagnose and prioritize problems and to initiate solutions by identifying and

mobilizing material, monetary, informational, technological, and human resources.

4. Finally, *rapidity* is the capacity to restore functionality in a timely way, thus containing losses and avoiding disruptions (Tierney and Bruneau 2007: 15). ●

The concept of self-organized criticality has been applied to ecosystems, economies, and urban systems. Gisiger (2001) reviewed evidence that ecosystems self-organize into a critical state as the web of interactions among species and individuals develops. According to Krugman (1996), self-organized criticality might explain the evolution of contemporary cities. Krugman identified two principles of self-organization. The first is the emergence of "order from instability," which is evident in the formation of Los Angeles–style "edge cities." A second principle that characterizes urban evolution is the emergence of "order from random growth." Krugman showed that the distribution of urban populations evolves to a critical state that is represented by Zipf's law: the number of people in a city is inversely proportional to its rank among all cities. Urban scholars have hypothesized that cities reach a critical point that explains the evolution of various urban phenomena such as urban form (e.g., urban development in Buffalo and western New York; Batty and Xie 1999) and collective behaviors (e.g., traffic fluctuations in London; Petri et al. 2013). It is the self-organized criticality of hybrid systems that provides opportunities for resilience and adaptation and the capacity for innovation—they are emergent properties of the systems' complex dynamics.

Resilience and Innovation as Emergent Properties

Studies of complex systems reveal that systems that are *heterogeneous* and *modular* tend to be better able to adapt than those whose elements are highly connected and homogeneous (Scheffer et al. 2012). Other properties of complex systems that enhance adaptive capacity and innovation are *cross-scale interactions*, *early warning mechanisms*, and *self-organization* (Walker et al. 2004). Heterogeneity facilitates *flexibility* and lets a system function under a wide range of conditions (e.g., multimode transportation). When Hurricane Sandy flooded and shut down portions of the subway in New York City, the flexibility inherent in the imperfect, multimode, and redundant urban infrastructure provided alternatives and paths around areas of failure.

Modularity allows *autonomous functionality* and lets a system contain disturbances and avoid cascading effects. The electrical transmission grid is an example of an interdependent network. Failure in one place can lead to failure in others, as in the Northeast blackout of 2003, which impacted 55 million people. Connectivity stabilizes some processes but may allow a failure at one point to propagate across, and involve, an entire network. In contrast, modularity within electric grids allows technicians to isolate failures.

Self-organization is a process in which patterns at the top of a system emerge from interactions among lower-level components. Self-organization enables natural and social systems to change their internal structures and functions in response to external circumstances. These novel, emergent forms of social organization and socioecological function generate opportunities for transformation. Cross-scale interactions allow for functional redundancy across scales and permit a system to respond to disturbance through service substitutions. Just as the species of a functional group can develop a resilience that operates across scales, so, too, can political, economic, and urban infrastructure systems operate at multiple scales to reinforce system functionality.

Regime Shifts in Urban Ecosystems

Cities face unprecedented challenges. Global environmental change is placing increasing pressure on ecosystem functions and their ability to support human activities. The exponential growth of human activities is a key driver of such change—so much so, in fact, that Planet Earth has clearly entered a new epoch: the Anthropocene, in which humans have as much influence as nature itself (figure 4.1; Steffen, Crutzen, and McNeill 2007). Urbanization is a key driver of global-scale phenomena—such as climate change—that are threatening the ecosystem's capacity to deliver important ecological services (Alberti 2010). At current rates of urban growth, expected changes in global land cover will result in

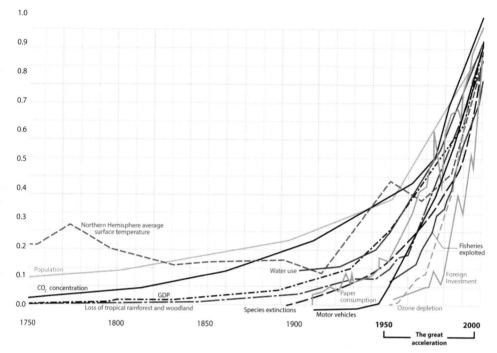

FIGURE 4.1 The Anthropocene. The term suggests that the Earth has now entered a new epoch. Over the past fifty years, human activities have accelerated a range of key trajectories, as is clearly visible in the twelve indicators charted between 1750 and 2000. The y-axis is a normalized index representing rates of change in human activities since the beginning of the Industrial Revolution. Some activities were not present before 1950; for others, the rate of change increased sharply after 1950. Source: Steffen, Crutzen, and McNeill 2007.

significant loss of habitats in key biodiversity hotspots (Seto, Güneralp, and Hutyra 2012). Urban regions are also where the majority of the human population will face the consequences of irreversible changes in climatic, hydrological, and ecological regimes, such as flooding, droughts, and sea-level rises (figure 4.2).

Rapid modifications of biophysical systems have the potential to trigger regime shifts: abrupt and irreversible changes with significant consequences for human health, safety, access to resources, and overall well-being (Rockström et al. 2009). Researchers have found that the likelihood of regime shifts is higher in ecosystems in which resilience has been reduced by humans via modification of biogeochemical cycles,

Flooding   Landslides   Drought   Earthquakes   Wildfires

FIGURE 4.2   Extreme events. Photographs: aerial photo of Hurricane Katrina flood: National Oceanic and Atmospheric Administration; landslide: European Pressphoto Agency; drought: Benjamin Jakabek; earthquake: Martin Luff; Zaca wildfire: John Newman, U.S. Forest Service.

alteration of hydrological regimes, reduction of biodiversity, and changes to the magnitude, frequency, and duration of disturbance regimes (Folke et al. 2004). Potential drivers of regime shifts—from climate change and flooding to water pollution—pose enormous challenges to the stability of urbanizing regions and render them vulnerable (A. W. Miller et al. 2009). Hurricane Sandy, in 2012, and the less recent but devastating Hurricane Katrina, in 2005, along with the 2011 Tōhoku tsunami, are early examples of the shocks that coastal cities will face in coming decades. How are cities to navigate such uncertainty and make robust decisions to ensure human well-being over the long term?

Regime shifts are large, abrupt changes in the structure and function of a system (as illustrated by the simple graph in figure 4.3a (cf. figure 2.11,

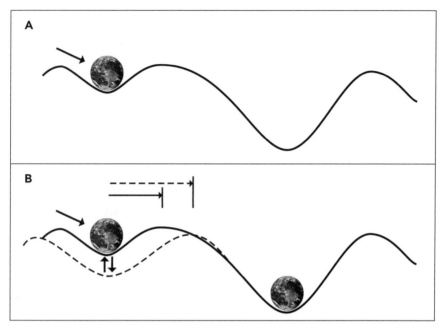

FIGURE 4.3a–b   (a) Regime shifts are large, abrupt, persistent changes in the structure and function of a system. An external shock can trigger a completely different system behavior, here represented by the ball moving into a new regime. (b) But regime shifts also depend on slow changes in external drivers and internal feedbacks that change the regime's domains of attraction: from a resilient state (indicated by the dotted line) to a less resilient one (indicated by the solid line). A state's resilience corresponds to the width of a stability pit. Loss of system resilience changes the thresholds that push the system into a new regime. Definition of regime shifts: Biggs et al. 2012. Photograph: Earth: woodleywonderworks.

in chapter 2). A *regime* denotes the behavior of a system, maintained by mutually reinforced processes or feedbacks, within a stability domain represented by a "valley," or basin of attraction, in the stability landscape (figure 4.3a). Regime shifts are shifts in dominant feedback maintaining the system dynamic. An external shock, such as an extreme climatic event, can trigger a completely different system behavior (figure 4.3b). Regime shifts can also lead to hysteresis: the path of recovery or return (e.g., behavior or time scale) to the original state may differ from the degradation phase that led to the transition.

Scientists have documented several examples of regime shifts in eco-

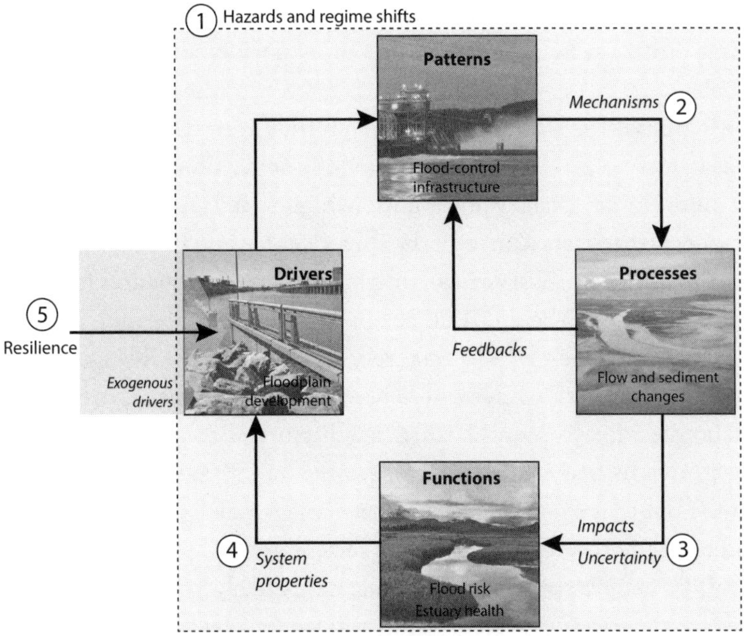

FIGURE 4.4 Applying the resilience framework to flood regimes in urban ecosystems. Hydrological regime shifts may emerge from interactions between flood-control strategies and hydrological processes. The flow and sediment changes imposed by flood-control infrastructure (FCI: channelization, levees, and flood-control dams) prompt river adjustment—further geomorphic changes—that eventually affect flood risks by altering the ecosystem's hydrological resilience. Flood control and erosion simultaneously affect ecosystem health and human risk, prompting increase in flood-control infrastructure. Photographs: floodplain development: Lisa Campeau; flood-control infrastructure: Mark Lehigh; flow and sediment change: Evan Leeson; flood risk and estuary health: DCSL.

systems: a coral reef can become an algae reef, a tropical forest can be replaced by grassland, and clear water can become eutrophic (Stockholm Resilience Centre 2015). More recently, they have described several examples in urbanizing regions (i.e., urban lakes, invasive species, floods), but we still do not fully understand the significance of such dynamics. In urbanizing regions, they emerge from complex interactions among natural processes and human interventions, which modify feedback mechanisms that maintain system functions and prevent the system from shifting to alternate states (figure 4.4). For example, interactions between human and natural drivers in coastal cities, may generate regime shifts, such as flooding, loss of estuary biodiversity, and algal blooms.

All around the world, climate events that are extreme compared to those in historical records are becoming the norm. Climate scientists predict more hot days, heavy precipitation, high-speed winds, and increased numbers of hurricanes (in, e.g., the United States and the Caribbean) and flash floods (in, e.g., East Africa), with significant consequences for human and ecological well-being (IPCC 2012). For the Intergovernmental Panel on Climate Change's 2012 special report, *Managing the Risks of Extreme Events and Disasters to Advance Climate Change Adaptation*, 220 scientists collaborated for 30 months to review historical trends and projected trajectories to assess current frameworks and strategies. Their report pointed out that existing measures to manage risk and adapt to change need to be improved dramatically to face projected climatic extremes.

By 2030, storm surges and sea-level rise are likely to force many coastal cities to contend with waters rising at least four feet (about 1.2 meters) above current high-tide lines (Climate Central, Inc. 2012). Strauss et al. (2012) estimated that urban areas and their communities are highly vulnerable to sea-level rises of up to six meters above mean high tides. In the United States alone, 22.9 million Americans live on land within six meters of local mean high tides. Historically, river deltas would shift location and expand and contract—depending on changes in precipitation and erosion—within a stability domain. Slow changes in internal feedback caused by hardening of banks and levees may change the domains of attraction from a resilient condition to a less resilient one, thus changing the thresholds that push the system into a new regime.

Interactions Leading to Socioecological Innovation

Critical transitions pose great challenges. At the same time, they offer unique opportunities for positive transformation and generate the seeds for cities to become more resilient and innovative. When new frames of reference and new constraints emerge, they require people and planners to recognize opportunities and risks they have never experienced before: to expand the boundary conditions of what is possible and desirable.

I hypothesize that social-ecological innovation in urban and urbanizing systems emerges from interactions among human, natural, and technological processes. The generation and diffusion of novelty are part of a tightly coupled system of human and ecological drivers. For example, the generation and adoption of efficient technologies (e.g., energy, water, and CO_2 emissions) are driven by increasing social interactions (e.g., face-to-face interactions and collaborations) and increasing environmental changes (e.g., extreme climatic events). Technological innovation and diffusion might be further controlled by economic growth, urban development policies, and urban management strategies. The strength and dominance of these factors might also vary across cultures and biomes.

Uncertainty

At the core of the challenge we face is the inevitable uncertainty intrinsic to dynamic coupled human-natural systems (Liu et al. 2007a). Both ecosystems and societies experience continuous fluctuations in their structures and functions. Tidal phenomena provide an example of short-term natural fluctuations. Occasionally, the predictable continuous changes are punctuated by sharp shifts: abrupt transitions to alternative states with significant implications for system functions and dynamics (Scheffer et al. 2001, 2012).

Scientists have long documented examples of regime shifts (Rocha, Peterson, and Biggs 2015). Among those often observed in urban regions is freshwater eutrophication associated with excessive discharges of phosphorus from urban wastewater. The Lake Washington eutrophication phenomenon in Seattle is a well-documented example of a regime shift driven by a large amount of nutrients delivered by a rapid increase in urban sewage and simultaneous widespread use of phosphate-based detergents during the 1950s. It took a public vote and the diversion of sewage from the lake into Puget Sound (completed in the late 1960s) to improve water clarity and restore fish habitat (Edmonson 1991).

Strategic decisions about urban infrastructure and growth management are based on our assessment of the past and our expectations for the future. But our predictions can account for only some interactions between highly uncertain drivers of change and surprising, but plausible, futures over the long term. Important progress has been made in predictive modeling, and improved simulation and computer power have allowed us to process astonishing amounts of data. Still, our models are constrained by our limited knowledge, unverified assumptions, and short-sighted mindsets.

Change is the way that living systems work, but most ecological design for urban development is based on a myth: it assumes that we should aim to produce policies and developments that result in social, economic, and environmental stability. Yet novelty and innovation are essential elements if systems are to maintain their functions and resilience over time (C. R. Allen and Holling 2008). A more appropriate strategy entails adopting a dynamic approach that embodies change and discontinuity.

A Framework for Urban Ecology

Understanding relationships between urbanization and ecosystem function is critical to generating a theory of resilience and innovation in urbanizing ecosystems (Alberti and Marzluff 2004). Strategic decisions concerning urban planning and investments in infrastructure require syntheses of complex and evolving knowledge of the workings of coupled human-natural systems and their sustainability, resilience, and transformation. Only through new productive collaboration among researchers from a broad range of disciplines can such syntheses address questions like: Are there general properties of cities' socioecological structures and governances that predict their capacities to adapt, innovate, and transform?

To study urban regions as complex coupled human-natural systems, we must build on various theoretical frameworks. Complex systems theory provides a powerful approach to link urban sustainability and resilience by explicitly modeling interactions and feedbacks between human and natural processes in urban and urbanizing ecosystems. Using this approach, we can aim to identify the properties of a city's socioecological structure and governance that predict its resilience and innovative potential. Developing and testing hypotheses linking urban structure and dynamics to resilience require a novel approach able to integrate multiple methods (F. Miller et al. 2010) and address methodological challenges inherent in the study of urban ecosystems.

Here I propose four questions:

1. How do socioecological interactions and feedbacks affect adaptability, resilience, and innovation in urban ecosystems?
2. How do these interactions vary with patterns of urbanization (e.g., urban size, density, heterogeneity, and modularity)?
3. How do emergent trajectories feed back into patterns of urbanization, economic development, and ecosystem and human well-being?
4. How can planners and policymakers integrate this new knowledge base to create sustainability and resilience in urbanizing regions?

A novel framework to study innovation in urban socioecological systems must consider economic and biogeophysical factors simultaneously. Using such a framework, we can begin to characterize the couplings and feedbacks among human, natural, and technological systems as emergent properties of interactions between urbanization and environmental change, and we can examine their roles in generating novelty and promoting the diffusion of innovation. Present cities function as natural laboratories for understanding generative socioecological innovation processes and represent hypothetical microcosms of alternative futures for an urbanizing planet.

A Research Agenda

1. How do socioecological interactions and feedbacks affect adaptability, resilience, and innovation in urban ecosystems?

Studies of complex systems reveal that heterogeneous and modular systems tend to have greater capacities to adapt than do highly connected and homogeneous systems (Sheffer et al. 2012). Other properties of complex systems that enhance adaptive capacity and innovation are cross-scale interactions, early warning mechanisms, and self-organization. To advance a theory of resilience in urban ecosystems, we must determine which, if any, of these properties of resilient structures apply to complex urban ecosystems. Using cross-city comparisons, researchers can refine hypotheses about specific structural components, the dynamics of resilient urban systems, and how they differ across cities with different geographic, socioeconomic, and ecological settings and habitation histories. Key questions for a research agenda include the following:

- Which characteristics of urban systems are able to evolve instead of maintaining and reinforcing the status quo? Do urban systems have identifiable properties that predict whether these systems will be resilient and robust?
- Which patterns of urbanization and infrastructure influence the mechanisms that underlie scaling relationships of urban agglomerations? That is, what are the dynamics of social interactions and innovation?
- What roles do the spatial concentration and temporal acceleration of social interactions play in governing system resilience and innovation (Bettencourt 2013)?
- As cities' decision-makers, planners, and stakeholders face global change, are signals available to inform them about the relative benefits of major transformations versus those of simply bolstering existing capacities for resilience?
- Do common elements exist within the structures of urban systems and their governing institutions that will allow people to successfully navigate climate change–driven socioecological problems in a sustainable way? If so, can these elements be generalized to broader contexts and applied to cities worldwide?

Many studies indicate that the likelihood of regime shifts may increase when humans reduce ecosystem resilience by modifying biogeochemical and hydrological cycles, by reducing biodiversity, and by changing natural disturbance regimes. Many studies have addressed the relationship between urbanization and environmental change; few, however, have examined the effects of interacting human and ecological patterns and processes on cities' vulnerabilities and resilience to climate change. We do not know, for example, which socioecological tradeoffs are associated with variations in infrastructure and housing density in coastal cities, or how such variations impact human and ecosystem well-being.

2. How do these interactions vary with patterns of urbanization (e.g., urban size, density, heterogeneity, and modularity)?

Relationships between social and environmental scaling and socio-ecological innovation in urbanizing regions might be driven by differences in the relative importance of human or biophysical drivers, depending on varying biogeographic socioeconomic, and historical conditions. For example, increasing social interactions in cities might drive invention of new technologies, while environmental change could be a more significant driver of diffusion of such new technologies in cities that are highly vulnerable to extreme climatic events such as hurricanes, floods, heat waves, or persistent droughts. Through comparative studies of cities across different bioregions, we can determine the conditions under which biophysical drivers dominate social drivers of socioecological innovations.

The study of complex networks provides a powerful framework for investigating mechanisms governing these relationships. As I mentioned earlier, recent studies of complex networks reveal that systems with greater heterogeneity and modularity tend to have greater adaptive and innovative capacities than those characterized by homogeneous elements. Heterogeneity can allow redundancies that increase robustness to structural disruptions (although an excess of diversity might weaken redundancy). Heterogeneity is seen as a proxy for flexibility in diversification of functional roles and is enhanced by modular structures that allow functional autonomy (Newman 2006). While these network properties have been documented for unipartite networks, urban socioecological systems are far more complex. Urban systems are characterized by hierarchical, coupled (e.g., social, economic, infrastructure, and water and biogeochemical fluxes), and hybrid networks. Such hybrid networks might exhibit qualitatively different structural and dynamic properties governing the function of urban systems.

3. How do emergent innovation trajectories feed back into patterns of urbanization, economic development, and ecosystem and human well-being?

To navigate the transition to a sustainable urban future, it is necessary that we understand cities as integrated social, economic, and physical systems in more precise and predictive ways. This requires quantitative models of the internal structures of cities and of the interactions between cities and the Earth's natural environments that account for the processes of human development and economic growth, as well as for their feedbacks on patterns of urban development. Emerging patterns of socioecological innovation might indicate a trajectory for decoupling urban growth from consumption and might provide insights for economic investments and planning decisions that can simultaneously improve ecosystem and human well-being. Using scenario analyses and simulation modeling, we can explore the impact of emerging innovations on future patterns of urbanization.

4. How can planners and policymakers integrate this new knowledge base to create sustainability and resilience in urbanizing regions?

Over the past decade, resilience, adaptation, and reduction of vulnerability have become the centerpieces of dozens of national and international municipal plans. While empirical research has demonstrated that maintaining resilience is critical to socioecological well-being (Walker et al. 2004), few studies have tied specific policies and institutional settings to increased resilience. Effective planning and policy-making require the synthesis of emerging knowledge on coupled human-natural systems, bringing institutional actors and processes into ecological resilience theory. Furthermore, efforts to effectively integrate uncertainty into the decision-making process are challenged by the dynamics of sociocultural values, power domains, and knowledge. New tools such as strategic foresight and scenario planning are critical for supporting robust decisions under a broad range of diverse yet plausible futures.

Expanding the Time Scale of Decision-Making

Many problems posed by environmental (e.g., climate) change are already detectable. Others may not become apparent for many decades. Yet to continue business as usual, in light of what we already know and can anticipate, could result in catastrophic outcomes for urban ecosystems as they approach critical thresholds and regime shifts. Building adaptive capacity and resilience to environmental change into new system states is not a short-term process; it requires communities to rethink the time scales of their decision-making.

- Can historical and contemporary examples of effective long-term planning provide insights into what allows such systems to come into existence, to generate effective long-term plans, and to persist for extended periods of time?
- Can scientists and practitioners produce knowledge, models, and data that help decision-makers and stakeholders increase the spans of time they consider in their planning?
- Which institutional (formal and informal) changes may be necessary to facilitate movement toward long-term thinking in decision-making?

Resilience and Innovation: Toward Sustainable Institutions

Efforts are underway, led by multiple stakeholders, to develop strategies for adapting to expected climate change. Agencies across the world are setting targets, developing policies, and funding development and infrastructure projects; their success hinges on interactions between uncertain elements of future climate change and institutional responses. Scientific uncertainty increases the likelihood of political conflict among stakeholders about the most appropriate climate-change adaptation measures to implement. Uncertainty about climate-change risks, combined with differing risk perceptions and values among stakeholders, hampers local officials' efforts to reach decisions and implement adaptation plans.

How, then, can state agencies facilitate the efforts of city governments to implement plans to adapt to climate change? Which stakeholder processes can best support the development of alternative institutional designs to facilitate effective adaptation? What are the key players, conflicts, and institutional dynamics governing adaptations to climate change that can reduce vulnerability and increase resilience to extreme events (e.g., flooding and water resources scarcity)? How can processes of institutional innovation (cultural, institutional, and technological) be harnessed to help coastal cities become sustainable in the face of regime shifts induced by climate change?

Knowledge Synthesis

Resilience theory offers a robust framework for understanding what triggers regime shifts and the alternative pathways through which ecosystems and societies navigate change. Understanding regime shifts in urban and urbanizing systems will require new collaborations across the social and ecological sciences to identify key processes and mechanisms governing their dynamics.

A critical first step is to integrate human and ecological functions and well-being into current frameworks as key drivers of alternative ecosystem states. To integrate human and ecological processes in resilience science, we must not only acknowledge and expand the agents and mechanisms governing their dynamics; we must also redefine system boundaries, scales, and reference conditions.

Then, building on the fields of strategic foresight and stochastic modeling, we can frame resilience assessments within the context of multiple plausible future conditions. Instead of trying to eliminate uncertainty, we would aim to characterize uncertainty and incorporate it into the decision-making process in order to support robust management strategies (Peterson, Cumming, and Carpenter 2003). Scenario planning presumes that, in highly uncertain conditions, there is no single "optimal strategy" but rather a set of possible strategies that will allow a city to prepare for an array of potentially divergent futures.

We also need to identify indicators of resilience. Regime shifts in coupled human-natural systems are difficult to predict. Systems may reach critical tipping points before they exhibit recognizable change. Insights into navigating such uncertainty come from a broad spectrum of research disciplines. At one end are studies that focus on the architectures and dynamics of complex networks that enable resilience and transformation. On the other end are studies that aim to uncover generic properties of resilience and empirical indicators of proximity to critical thresholds (Bettencourt 2013; Sheffer et al. 2012). Indicators provide common foci on which the various efforts of scientists, practitioners, and policymakers converge. Properly developed indicators incorporate scientific information to portray linkages between systems' socioecological and economic components and to provide means of analyzing trends. Indicators must be scale-specific, relevant, and responsive to the goals of the human component, providing vital feedback between the state of the system and its managers (Alberti, Russo, and Tenneson 2013).

A final element is coproduction of knowledge. The efficacy of an adaptation strategy relies on diverse social actors and the participation of their formal and informal institutions (Ostrom 2009). Strategic planning decisions for urban growth and investment in public infrastructure require syntheses of complex knowledge from a diverse array of experts. Communities must make difficult investment and development decisions in the face of substantial scientific uncertainty regarding the nature of risks. Such uncertainty likely will increase political conflict among stakeholders over potential adaptation measures (Jasanoff and Martello 2004). Complex metropolitan systems cannot be managed by a single set of top-down governmental policies (Innes and Booher 1999); instead, they require that multiple independent players coordinate their activities under locally diverse biophysical conditions and constraints, constantly adjusting their behaviors to maintain balance between human and ecological functions. A resilient institution must provide a neutral arena in which diverse communities are allowed to understand the various roles and interests of urban stakeholders, and where they are ensured access to the decision-making process.

5

Resilience in Hybrid Ecosystems

VARIABILITY AND CHANGE ARE VITAL TO THE PERSISTENCE AND EVO-lution of ecosystems. Studies of complex systems have uncovered direct rela-tionships between variability in systems' structures and their resilience. Change—both slow and fast—is integral to the workings of any natural system. This chapter articulates the hypothesis that variable patterns of urbaniza-tion and modular urban infrastructure may be key to cities' resilience. I use three examples—carbon, nitrogen, and bird diversity—to illustrate the com-plex relationships between patterns of development and key slow and fast variables that regulate resilience in urban ecosystems. I argue that policies and management that aim to achieve stable conditions by optimizing only one system function at one scale may make systems more vulnerable and could eventually lead to their collapse.

Resilience in hybrid ecosystems is an emergent property of co-evolving human and natural processes. We cannot understand the diverse expres-sions of present urban landscapes unless we consider the complex his-tory of interactions between humans and nature over millennia. The collapse of the Mayan cities during the eighth or ninth century—still quite an enigma—has only recently been investigated from the perspec-tive of coupled human-natural systems by integrating empirical evidence from different fields. A plausible hypothesis is that a severe drought, exacerbated by rapid deforestation and desertification in a time of unprecedented population density, led the Mayans to abandon urban sites in the Yucatán region (E. Cook, Hall, and Larson 2012). Complex interactions between human and natural processes are also key to understanding the resilience of old cities, including Rome and many others in the Mediterranean region, that have survived the test of evolving nature and civilizations. The landscape stratification that we can see in Rome today reveals how humans and nature co-evolved over nearly 2,800 years of change: ecological, economic, social, cultural, and political. To explain the structure, dynamics, and evolution of emerging ecologies in urban ecosystems—whether we are interested in the biodiversity of New York's Central Park or Moscow's Bitsevsky Park, or in the biogeochemistry of Seattle or Phoenix—we must acknowledge their hybrid nature.

Although scholars of urban ecology have recognized the evolutionary nature of urban ecosystems and acknowledged their unique hybrid dynamics for some time, most empirical research is still grounded in divided paradigms. The knowledge that emerges from these paradigms is incomplete in a fundamental way. Emerging studies of coupled human-natural systems reveal new and complex patterns and processes that are not evident when social or natural scientists study them separately (Liu et al. 2007a). For the past few decades, teams of biologists, earth scientists, economists, geographers, and planners have expanded our understanding of urban ecosystems by uncovering key mechanisms that characterize coupled human-natural dynamics in urban regions (Alberti et al. 2003; Grimm et al. 2000; Pickett et al. 2001).

Several important findings have emerged from this work (Alberti 2010; Grimm et al. 2008a; Pickett et al. 2011). Cities have a distinctive biogeochemistry because infrastructure that is engineered to move water and remove wastes alters hydrological processes and nutrient cycles (Kaye et al. 2006). Densely urbanized areas also have unique microclimates, such as heat islands, which influence atmospheric chemistry and air pollution (Grimm et al. 2008b). Evidence also suggests that the distinctive compositions of plant and animal species found in cities are strongly influenced by human perceptions and behaviors (Faeth et al. 2005). But these specific findings do not add up to an understanding of how such systems work and evolve; we must uncover the rules governing community assembly.

One critical aspect of such an understanding is learning how the structures of urban ecosystems (i.e., diversity of components and degree of connectivity) relate to their dynamics. That is, which qualities best express and regulate function and change in hybrid systems? Recent evolution in complex science has begun to uncover direct relationships between complex network structures and their resilience. Scheffer et al. (2012) noted that two key qualities of system architecture—*heterogeneity* and *modularity*—might determine the likelihood of critical transitions and the emergence of thresholds and system shifts (i.e., a catastrophic bifurcation).

Variability and change are two essential characteristics of ecosystems. It is the great variability found in nature that explains persistence and evolution. Change, whether gradual or abrupt, is integral to the way nature works. Emerging evidence shows that complex networks in which components vary and connectivity is incomplete tend to have greater capacities to adapt than those characterized by highly connected homogeneous elements. Interactions among components that are relatively independent allow them to change and evolve in an autonomous way.

In this chapter, I articulate the hypothesis that heterogeneity and modularity of urban structures may be key to cities' resilience: their ability to adapt to changes in ecosystems and in human communities. I challenge the dominant assumption of traditional urban planning that one optimal pattern of urbanization is consistently more resilient than another. I propose that policies and management that apply fixed rules for achieving predictable, stable conditions by optimizing one function at one scale may, in fact, make systems more vulnerable and eventually lead to their collapse. By simulating the effects of managing short-term variance in three models of ecosystem services—lake eutrophication, harvest of a wild population, and yield of domestic herbivores on a rangeland—Carpenter et al. (2015b) show how interventions to decrease variance might create ecosystem fragility by changing the boundaries of safe operating spaces and cancelling signals of declining resilience. A paradigm of co-evolution between humans and nature may prove more appropriate for learning how cities work and adapt to change.

Resilience in Hybrid Systems

It is critical to establish empirical evidence about the relationships among structure, dynamics, and resilience in urban ecosystems in order to advance sustainability science and effectively address emergent problems facing cities across the world. Holling (1973) defined *resilience* as the capacity of a system to absorb disturbance from change and reorganize while maintaining essentially the same functions, structure, identity, and feedbacks. Ecological resilience in urban ecosystems is governed by complex interactions between human and ecosystem functions over multiple scales and involving multiple heterogeneous agents (Alberti and Marzluff 2004; Peterson, Allen, and Holling 1998). We must include these complex interactions and agents when we study such systems. If we are to understand systems dynamics and potential shifts, it is not sufficient to consider human and ecosystem functions separately, because integrated socioeconomic and ecological systems behave differently than their separate parts. Furthermore, since urban development patterns affect the amount and pattern of built and natural land cover and the demand for natural resources to support human activities, alternative urban development patterns (e.g., urban form, land use distribution, and connectivity) may have differential effects on resilience. The challenge to effective planning and management of coupled human-natural systems is to expand our knowledge of their dynamics, resilience, and capacity to adapt and of their potential variability across regions and scales.

In 2004, John Marzluff and I described how, as ecosystems become urbanized, they move between two basins of attractions or stable equilibria: a natural state dominated by natural processes and an urbanized state dominated by humans (figure 5.1). As a region urbanizes, ecological processes supporting the urban ecosystem may reach a threshold and drive the system into an unstable state. Eventually the system shifts to a new state in which ecological processes are highly compromised and drive the system to collapse. We defined the resilience of an urbanizing region as the system's ability to maintain both types of functions simultaneously.

Urban rivers are an informative example of the evolving legacy of disturbance and system dynamics in urbanizing regions. Regime changes in river systems, and in their ecosystem functions associated with human action, have been described by several scholars looking at key geomorphological, hydrological, and ecological processes across a gradient of urbanization (C. J. Walsh et al. 2005; Vietz, Walsh, and Fletcher 2015; Hopkins et al. 2015) (figure 5.2). Over the past decade, we have learned a great deal about the complex interactions between changes in river functions and changes in land use and infrastructure.

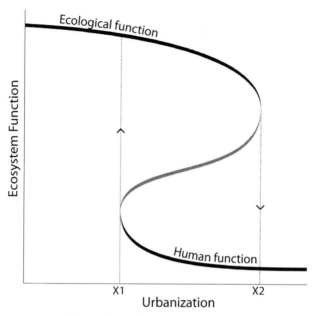

FIGURE 5.1 Dynamics in urban ecosystems. As a region urbanizes, ecological processes supporting the urban ecosystem may
reach a threshold (X2) and drive the system into an unstable
state. Eventually the system shifts to a new state in which ecological processes are highly compromised and drive the system
to collapse. Alberti and Marzluff 2004.

Through a variety of human activities, urbanization affects hydrology
and the pathways, rates, and frequency of water movements in a watershed, as well as associated sediment budgets and stream channels.
Changes in the physical template may interact with external changes
such as climate change and may drive regime shifts in the ecology of a
stream. Although, to date, the best-known examples of system shifts in
urban rivers involve geomorphic or hydrologic phenomena (Dent, Cumming, and Carpenter 2002), more recent studies have pointed out positive feedback mechanisms in stream ecology. In urban streams, riparian
and in-stream vegetation may play a critical role in creating and maintaining distinct alternate states that are characterized by markedly different ecological communities and processes (Heffernan, Sponseller, and
Fisher 2008; Stanley, Powers, and Lottig 2010).

Combined sewer overflow and outfalls; Waterfront

Underground pipes and lateral drains; CBD

Gutter and street drains; Residential neighborhood

Swales and retention ponds; Exurban development

Infiltration and natural waterways; Intact forest

FIGURE 5.2   Urban hydrology has implications for urban infrastructure across a gradient of urbanization. Maps (*left to right*): LiDAR Top Surface DEM 2003: King County, Washington (2003); infrastructure, sewer, and drainage: City of Seattle 2012; property/survey: building outlines (City of Seattle 2012), street/transportation: City of Seattle 2012. All photographs: Google Maps 2015.

Researchers have documented more and more examples of regime shifts (see the Regime Shifts Database, www.regimeshifts.org). More recently, they have described several examples in urbanizing regions (e.g., urban lakes, invasive species, and floods), but we still do not fully understand the significance of such dynamics. In particular, we do not know the extent to which patterns of urbanization might mediate relationships between urban stressors and ecosystem functions and resilience.

Resilient Urban Patterns: A Hypothesis

Increasing numbers of studies show that patterns of urbanization have differential effects on ecosystem functioning, and that those patterns may mediate ecosystem responses in subtle and unexpected ways (Bang, Sabo, and Faeth 2010; Faeth et al. 2005). The hypothesis that patterns of urbanization influence ecosystem function relies on the assumption that several thresholds exist (figure 5.3a). The dominant view in urban

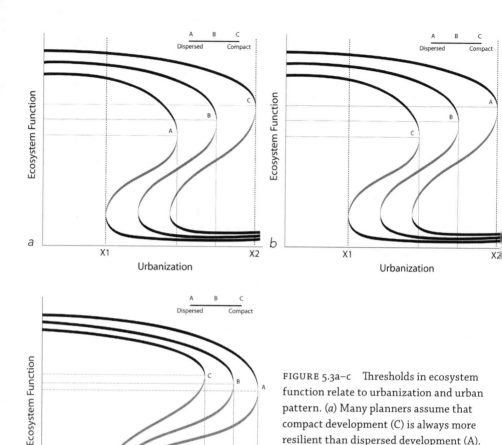

FIGURE 5.3a–c   Thresholds in ecosystem function relate to urbanization and urban pattern. (a) Many planners assume that compact development (C) is always more resilient than dispersed development (A). (b) However, for some ecosystem functions, dispersed development (A) can be more resilient than compact development (C). (c) Adaptation of ecological process to urbanization might also shift the threshold in system function, moving it to the lower right.

ecological planning is that dispersed development (A) leads to a faster decline in system function compared to more compact forms of development (C) and that, therefore, compact development is more resilient (figure 5.3a). But because patterns of development have different impacts on the slow and fast variables that regulate system dynamics, the reverse can also be true: dispersed development can be more resilient than compact development under certain conditions and for certain ecosystem functions or scales (figure 5.3b). Also, we do not know whether adap-

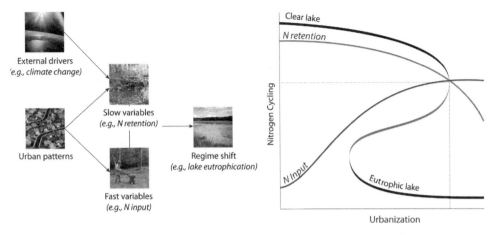

FIGURE 5.4 Urban patterns have complex impacts on *slow* and *fast* variables that control conditions in urban systems (e.g., lake eutrophication). Slow variables (e.g., N retention in urban watersheds) refer to factors that change gradually, in response to long-term processes, and constrain the ranges of responses of fast variables (e.g., nutrient inputs). Photographs: climate change: Michel; urban patterns: Sam Korson; N retention: Yortw; N input: CIFOR; lake eutrophication: Dave Blogg.

tation of ecological processes to urbanization would make an urban ecological system more resilient to urban stress. In figure 5.3c, this phenomenon is represented by the system transition thresholds moving to the lower right. The shifting thresholds depend on dynamics and tradeoffs that we do not fully understand; on variable human and environmental conditions; and ultimately, on future interactions among uncertain trajectories of key driving forces.

Urban patterns have complex impacts on what we define as *slow* and *fast variables* that affect system resilience. Slow variables are factors that change slowly in response to long-term processes (such as the growth of trees or the accrual of sediment) and that constrain the responses of fast variables, potentially generating or preventing a tipping point. Fast variables are factors that change rapidly and that managers find easier to measure. For example, the slow variables affecting nitrogen budgets in a watershed are factors that change gradually in response to long-term processes (such as nitrogen retention) and that constrain the responses of fast variables (such as nutrient input), generating a shift to a new state (such as lake eutrophication) (figure 5.4).

The question, then, is not simply whether patterns of urbanization matter. Rather, we should ask: If they matter, how do urbanization patterns influence key variables and mechanisms that affect different ecosystem functions? Three examples—carbon, nitrogen, and bird diversity—illustrate the complex relationships among patterns of development and key slow and fast variables that regulate ecological resilience. In using these examples, my intent is not to focus specifically on any individual pattern or function but to emphasize nonlinear relationships among mechanisms affecting fast and slow variables that control ecosystem dynamics along a gradient of urbanization.

Signatures of Hybrid Systems

I begin with the assumption that different patterns of development produce different landscape signatures (spatial and temporal changes in ecosystem processes) and that, in turn, these signatures influence patterns of urbanization (Alberti 2008). If we could build cities by replicating a single pattern for which we could determine per-capita land and emission budgets, perhaps we could find an optimum pattern for the urbanizing world. But cities do not conform to a single pattern. Instead, they exist along a complex dynamic gradient of evolving landscapes: what was the urban fringe in 1900 is today the dense urban core. The signatures of urban ecosystem functions vary along a hypothetical gradient of urbanization, ranging from the urban core (characterized primarily by urban redevelopment) to suburban and exurban areas (where rapid development is occurring) to rural and intact forest. Evidence shows that patterns of urbanization mediate these dynamics in subtle and unexpected ways.

Urban Carbon

Urban development directly and indirectly affects stocks of carbon (i.e., pools of carbon such as plants). It also affects carbon fluxes: exchanges between two different stocks, such as the transfer of carbon dioxide from the atmosphere to the biosphere via plant photosynthesis or in the opposite direction via combustion of organic matter. Land-cover change is only one of many processes linking urban development patterns to the carbon budget. Urban development typically involves an increase in the amount of impervious surface, which alters hydrology and reduces infiltration capacity. Impervious surfaces and human activities may also change the microclimate (Oke 1982). In addition, urbanization involves multiple pollution sources, including chemical inputs from industry, agriculture, and transportation. Finally, land-cover changes typically result in changes in the abundance and composition of plant species, which affects the rates at which carbon is assimilated.

Hypotheses about the variability of carbon stocks and fluxes along an urban gradient are grounded in mechanisms that are known to affect carbon stocks and fluxes (Canadell et al. 2003). Lucy Hutyra and I have identified five key mechanisms that affect change in carbon stocks and fluxes along a gradient of urbanization: land use, emissions, organic inputs, temperature, and nitrogen fertilization (Alberti and Hutyra 2013). Carbon fluxes include exchange processes that are both positive (uptake: photosynthesis and soil accretion) and negative (loss: respiration and emissions). Carbon fluxes typically respond and change on much shorter time scales (e.g., hours) than carbon stocks, which change as a result of long-term changes in fluxes on a time scale of years or longer.

Scientific knowledge of the shifts in water and life cycles occurring in urbanizing regions is grounded in nearly a century of research, but shifts in carbon cycles associated with urbanization—specifically the dynamics and trade-offs that control system variables—are far more recent and are not well documented. We do not know the magnitude of carbon stocks in urban areas. For example, in the Seattle metropolitan area, my research lab team found that the amount of live biomass aboveground far exceeds previous estimates of carbon stocks (Hutyra, Yoon, and Alberti 2011). Furthermore, nowhere is the uncertainty and complexity of linkages and thresholds more pervasive than in urban forests (Kaye et al. 2006; Pataki et al. 2006).

One confounding variable leading to these challenges is the ubiquitous relationship between carbon cycles across biomes (regions) and scales. In a temperate environment such as the Seattle metropolitan area, reduction in forest cover is a primary driver of carbon cycling (Hutyra, Yoon, and Alberti 2011). However, in desert environments such as the Phoenix metropolitan area, where irrigation has radically transformed urban vegetation from the presettlement vegetation, relationships between urban land use and ecosystem productivity are driven largely by human alterations to the availability of water. When legacies such as agriculture, transportation infrastructure, and geomorphology are combined with the uncertain trajectories of key driving forces such as technological innovation and climate change, they significantly reduce our capacity to predict the differential effects and mechanisms that will govern relationships between urban patterns and ecosystem functions.

The curves in figure 5.5 show effects on carbon stocks and fluxes per unit of biomass (y-axes) as they are expected to vary along an urban gradient (x-axes). Compared to the urban center, plants take up more carbon at the urban fringe and increasingly more in the exurban areas and forest end point in proportion to the total biomass per unit area. But emissions are also higher at the urban fringe because suburban residents drive more miles. Trade-offs may exist between stocks and fluxes. For example, in dispersed developments, households can maintain relatively more live biomass than they can in dense urban areas because land parcels are larger, but suburban residents may produce higher emissions of carbon dioxide because they commute farther.

Variations in the carbon budget across an urban gradient are a result of the gradient's influence on several mechanisms that govern the uptake, release, and storage of carbon. Hutyra and I hypothesized that stocks of carbon within vegetation will be higher as urban intensity decreases but that the changes will be nonlinear (Alberti and Hutyra 2013). We predicted that urban vegetative stocks of carbon will be lowest where development is most intense because green vegetation will be replaced with buildings and pavement. Carbon stocks in organic detritus are kept artificially low in urban areas because cities collect and remove leaves and debris, but we would expect the fluxes (input rates) to increase linearly across the urban-to-rural gradient (which is directly proportional to biomass and leaf area).

Carbon fluxes typically respond and change on much shorter time scales (e.g., hours) than do carbon stocks, which respond, on a time scale of years, to accumulated long-term changes in fluxes. We hypothesized that, per unit, more biomass carbon will be taken up in urban settings because of the favorable growing conditions: people will water, fertilize, and prune their plants, and they will replace native vegetation. And, in the most intensely urbanized areas, more carbon will be lost through the ecosystem's heterotrophic respiration (R_H) per unit of mass due to

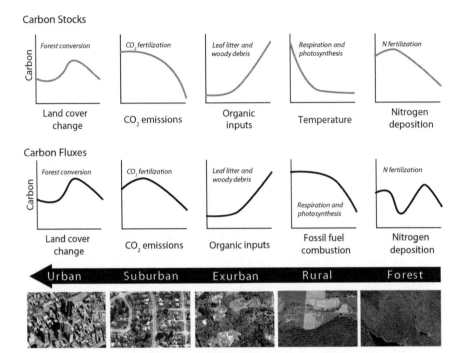

Carbon Stocks

Carbon Fluxes

Urban   Suburban   Exurban   Rural   Forest

FIGURE 5.5   Hypotheses of mechanisms influencing carbon budgets along the urban gradient, showing variability of carbon stocks and fluxes per unit of biomass (*y*-axes). They are expected to vary along an urban gradient (*x*-axes) in relation to key mechanisms governing carbon uptake, release, and storage (Alberti and Hutyra 2013). Aerial photographs: Google Maps 2015.

increased temperatures and soil moisture, but the removal of organic inputs (leaf litter and woody debris) will reduce the amount of substrate (stock) for decomposition and result in an overall decrease in R_H per unit of area. Higher levels of estimated CO_2 emissions are generally associated with the greater vehicle miles traveled (VMT) of suburban residents, but actual CO_2 emissions associated with travel have been observed both at and around the urban core. As concentrations of CO_2 increase, autotrophic respiration (R_A) can be expected to decrease. Higher urban concentrations of ozone will also dampen carbon uptake rates (gross primary productivity, or GPP), but increased atmospheric CO_2 concentrations will increase GPP.

Taken together, we hypothesize that these five mechanisms will

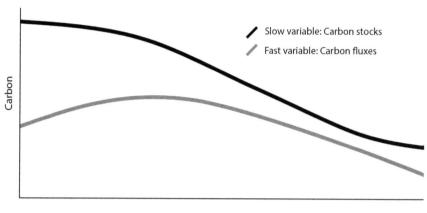

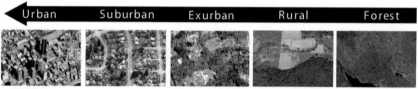

FIGURE 5.6   Relative impact of development on carbon stocks and fluxes. The slow variable (carbon stocks) is affected predominantly by dense development, while the fast variable (carbon emissions) is affected primarily by suburban development (Alberti and Hutyra 2013). Aerial photographs: Google Maps 2015.

produce nonlinear variations in carbon stocks and fluxes across the urban gradient (figure 5.6; Alberti and Hutyra 2013). The amount of carbon in vegetative biomass (and soils) is expected to generally increase with decreased development intensity, with a small peak in older suburbs and exurbs where larger lots have had time to accumulate biomass after they were initially cleared. Fluxes (per unit mass) might be expected to decrease with decreasing temperatures and decreased N and CO_2 fertilization, but, regardless of temperature, they ultimately will be highest in the least dense areas because of forests' high density of photosynthetically active vegetation.

Urban Nitrogen

Patterns of urbanization also affect the nitrogen budget in complex ways, through both changes in nitrogen input from multiple sources and changes in nitrogen retention. Nitrogen retention is affected by soil erosion and storm water, as well as by the extent to which nitrogen is present in a mobile form (e.g., fertilizer). It also can be taken up into biomass or denitrified, processes that depend on the temperature and available carbon. Nitrogen inputs depend on a combination of human activities that involve wastewater, fertilizer, and atmospheric NOx (figure 5.7).

The transition from forested or vegetated landscape to an urban landscape changes the relative contributions of these sources: from primarily atmospheric to wastewater, fertilizer, and possibly vehicle emissions inputs. Onsite septic tanks are found in more densely populated settings as suburban areas expand into the rural fringe. Higher densities of septic tanks can increase the levels of groundwater and of stream flow and, with them, nutrient concentrations (Sherlock et al. 2002). Impervious surfaces in urban areas are important pathways for accumulation of nitrogen, which is readily transported to water bodies (Collins, Sitch, and Boucher 2010).

Several mechanisms controlling soil's nitrogen content are potentially influenced by development, land-use legacies, and household landscaping practices: inputs (both intentional and unintentional), removal (of litter and debris), denitrification (conversion of nitrate to gaseous forms of nitrogen, which then diffuse back into the atmosphere), and movement of water out of yards (either through runoff to streams or infiltration to groundwater). Although residential developments in more urbanized areas tend to have high rates of nitrogen loading because residents apply fertilizer and deposit byproducts of combustion, people are also likely to remove litter and debris and convert land to impervious surfaces, activities that increase the amount and velocity of runoff. All this can result in loss of nitrogen. Residential parcels that were once farmland are likely to have comparatively high levels of soil nitrogen because of agricultural practices (Kaye et al. 2006). Household landscaping practices also frequently add nitrogen to the soil in the form of compost and synthetic fertilizers. On the other hand, removal of litter and debris may reduce nitrogen inputs, and frequent irrigation may enhance rates of denitrification and infiltration (Groffman et al. 2009).

I propose that the fast variable (nitrogen input) is affected primarily at the urban fringe due to the combination of nitrogen input from atmospheric deposition and fertilizer, while the slow variable (nitrogen reten-

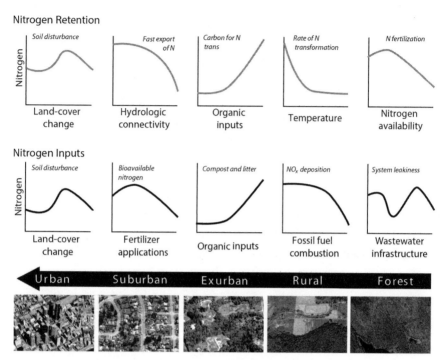

Nitrogen Retention

Soil disturbance	Fast export of N	Carbon for N trans	Rate of N transformation	N fertilization
Land-cover change	Hydrologic connectivity	Organic inputs	Temperature	Nitrogen availability

Nitrogen Inputs

Soil disturbance	Bioavailable nitrogen	Compost and litter	NOx deposition	System leakiness
Land-cover change	Fertilizer applications	Organic inputs	Fossil fuel combustion	Wastewater infrastructure

Urban   Suburban   Exurban   Rural   Forest

FIGURE 5.7   Hypotheses of mechanisms influencing nitrogen budgets along the urban gradient: The effects of urbanization on N budgets are mediated by complex interactions among fast and slow changes in N retention (represented by the upper set of curves) and N inputs (represented by the lower set of curves). Alberti and Larson 2011, personal communication; aerial photographs: Google Maps 2015.

tion) is significantly impacted at the urban core through highly connected impervious surfaces and pipes (figure 5.8).

Bird Diversity

Another example of this complexity is variability in bird diversity along an urban gradient, which involves the combined effects of loss of bird habitats (loss of forest cover and connectivity) and changes in biotic interactions, which lead to novel competitions. Complex interactions may emerge as habitats for native birds declines and humans create substitute habitats. Then new types of competition arise, resulting from the loss of native species coupled with colonization by early successional and synanthropic (thriving in human-altered habitats)

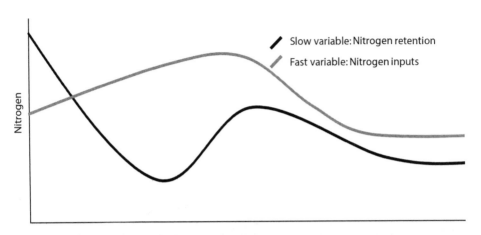

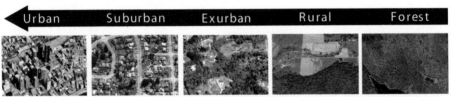

FIGURE 5.8 Relative impacts of development on nitrogen input and retention. Atmospheric deposition and fertilizer runoff affect nitrogen input (fast variable) primarily at the urban fringe, while at the urban core, N retention (slow variable) is mediated by connected impervious surfaces and pipes. Alberti and Larson 2011, personal communication; aerial photographs: Google Maps 2015.

species, which together generate a spike in the curve at the urban fringe (Marzluff 2005).

John Marzluff and I have developed a set of hypotheses to describe the complex interactions governing bird diversity across a gradient of urbanization. Figure 5.9 represents the hypothesized effects in relation to each mechanism as they vary along the gradient represented on the x-axis (Alberti and Marzluff 2013, personal communication). The y-axes vary according to mechanism. Rates of forest conversion and loss of native bird habitat are highest at the urban fringe, as forest conversion at the urban core has already occurred in the past. Forest connectivity declines as we move closer to the urban core. Food abundance is kept artificially high at the urban fringe because humans feed birds, but overall, it declines closer to the urban core. Disturbances increase with urbanization; we expect an increase in human-induced disturbances at and around the urban core.

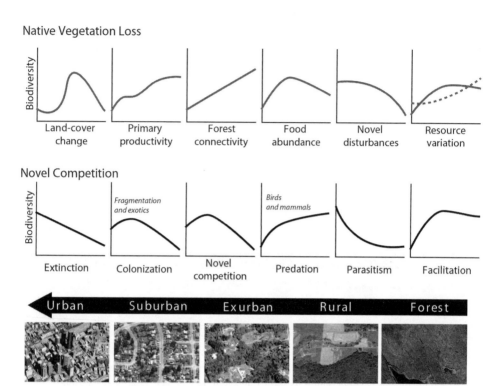

FIGURE 5.9   Bird diversity along urban gradients is governed primarily by impacts on natural habitat (fast variables) and biotic interactions (slow variables), though human supplementation of food supply can also play a role. (Alberti and Marzluff 2011, personal communication; aerial photographs: Google Maps 2015.

We can also expect different levels of variations in resources. Inputs of resources are replaced by human activities, and variation in resources may peak at the urban fringe, simultaneously reducing the variability of available resources; an example of this is the temporal variation in availability of food resources associated with rainfall patterns. Urbanization also alters mechanisms related to biotic interactions (Marzluff 2005). We expect a steady decline in native bird populations and an increase in extinction toward the urban core, while colonization by early successional and synanthropic species peaks at the urban fringe and then declines at the urban core. Predation from large mammals is expected to decline toward the urban core, but we do not expect a steady decline because urban pets also engage in a relatively substantial amount of predation (Marzluff

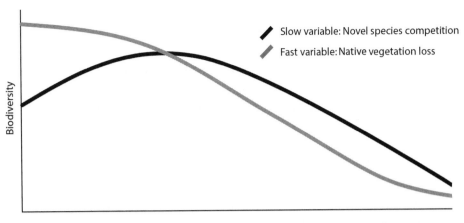

FIGURE 5.10   Relative impact of development on bird diversity. The fast variable (loss of native vegetation) is affected predominantly at the core, while the slow variable (increase in novel competition) is affected primarily by suburban development at the fringe. Alberti and Marzluff 2011, personal communication; aerial photographs: Google Maps 2015.

et al. 2007). We expect higher parasitism closer to the urban core. Humans may facilitate parasites in the suburbs (e.g., by putting out nest boxes). Insect species that are facilitated in "nature" by woodpeckers are facilitated in suburbs by woodpeckers, other birds, and people—for example, chickadees and swallows nest in nest boxes and lights. As diversity drops, some facilitators (e.g., large woodpeckers) might also drop out as their habitat is reduced and degraded. But we expect that this relationship will not be linear due to suburban activities; more likely, novel competitions will emerge as landscapes urbanize (Marzluff 2008).

Bringing these ideas together, I now highlight the nonlinearity (and perhaps the trade-offs), with respect to mechanisms affecting bird diversity, that exists across a gradient of urbanization (figure 5.10). Marzluff (2005) has found that, with increasing urbanization, changes in native bird habitat decline overall and that the relationships between bird diversity and the urban gradient do not adhere to a straight linear trend. At the urban fringe, the decline is less severe due to favorable growing conditions (e.g., human watering, fertilizer, pruning, and replacement of vegetation) and higher temperatures, but the decline increases with lower temperatures toward the fringe. Novel competition is higher farther from the urban core and peaks at the fringe because of subsidies and the relatively lower intensity of urban development. Of course, these mechanisms vary across biomes. We also hypothesize that trade-offs exist between colonization by early successional and synanthropic species and retention of native species; these mechanisms control the overall species richness of different patterns of development.

Human Perceptions and Behaviors

Human preferences and behaviors also drive changes in ecosystem function (figure 5.11). Biogeochemical cycling and biodiversity in urban areas are affected by human decisions about where to locate homes and businesses and how to manage the landscape. Carbon budgets and avian community compositions in urban regions are the result of complex interactions between human decisions and natural processes. Human decisions have selective implications for species diversity and provide supplementary resources in urban areas as people water their gardens and feed birds. In return, biogeochemical cycling and biodiversity provide important ecosystem services and amenities to urban dwellers. Most often documented are the impacts of urbanization on land cover and the associated loss of vegetated cover, both of which have significant consequences for carbon and wildlife. But human perceptions and behaviors also modify ecosystem processes in more subtle ways.

Humans provide supplementary resources in urban areas by watering gardens and feeding birds, both directly and indirectly. Vegetated cover simultaneously provides habitat for birds, sequesters carbon, and serves as an amenity for urban residents. It is well known that the incremental value of vegetation increases as resources become scarce, but we are only beginning to fully account for the individual and collective benefits of vegetated cover and environmental quality for properties and neighborhoods.

Many aspects of urban form may influence the amount of carbon that

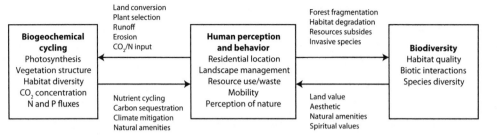

FIGURE 5.11   Human behavior. The fast variable (loss of native habitat) is affected predominantly at the core, while the slow variable (increase in novel competition) is affected primarily by suburban development at the fringe.

is emitted in urban areas (Pataki et al. 2006). Land can be developed in a wide array of patterns, based on variables such as form, density, land-use mix, and connectivity, and each pattern produces different rates of carbon emissions. A common assumption is that because suburban residents use their cars more than urban residents, they generate more vehicle miles traveled and therefore more emissions. But that assumption may not hold as urban regions become less monocentric in their development and job centers migrate beyond the urban core in response to increasing costs. Lucy Hutyra and I developed testable hypotheses about the carbon consequences of urban development by exploring linkages between urban form and carbon fluxes and by estimating the relative impact that alternative development patterns have on carbon fluxes across a gradient of urbanization (Alberti and Hutyra 2013). The impacts of carbon policies on residential and mobility patterns are difficult to detect because such policies are novel, and trends have yet to be observed over the long term. We expect that the complex dynamics between human perception and behaviors will generate nonlinear relationships between urban development and the key slow and fast variables that control the resilience of ecosystem functions.

Patterns of Resilience

Depending on alternative patterns, the trajectory of slow variables may constrain the impact of fast variables, either earlier or later in the process of urban development, by shifting the threshold associated with maintaining ecosystem function. This process is highly unpredictable and varies considerably under different bio-geophysical and socioeconomic conditions. It also depends on both ecosystem function (climate mitigation, biodiversity, biogeochemical cycling, etc.) and scale.

To achieve predictable and stable conditions, planning and management strategies tend to favor options that decrease variability of ecosystem processes over short time scales. Yet increasing evidence indicates that actions to decrease variance in ecosystems over short time scales may lead to long-term decline in ecosystem function (Carpenter et al. 2015b). Reducing short-term variance in ecological systems increases their variance in the long term. Evidence of how such phenomena, known as Bode's law, emerge in linear systems has been provided both in regulatory networks for ecosystem services (Anderies et al. 2013) and economics (Brock, Durlauf, and Rondina 2013). Along the same line, as I mention above, Carpenter et al. (2015b) have recently shown that reducing variance in nonlinear systems causes impact distributions to shift toward lower frequencies, leading to dynamical behavior changes and the risk of critical transitions.

In systems with multiple regime shifts, such as coupled socioecological systems, if one specific subsystem remains resilient at a specific scale, it may cause the entire system to lose its resilience in other ways. Urban mobility can be made highly efficient at the local scale by optimizing one type of transportation infrastructure, such as road transportation, while disinvesting in regional rail systems. A perturbation such as an unexpected shortage in gasoline or a spike in oil prices can result in system failure, however, simultaneously paralyzing multiple metro regions. The highly optimized tolerance (HOT) theory (Carlson and Doyle 2000) shows how due to robustness trade-offs, systems that are highly robust to frequent disturbances might become vulnerable to infrequent ones (Folke et al. 2010). If systems are to remain resilient, they must retain their adaptive capacity and their ability to cope with uncertainty across scales.

Predictability, Uncertainty, and Surprise

How do we deal with uncertainty and surprise? Consider, again, the hypothetical futures I described in the first chapter of this book, and imagine

Climate Change

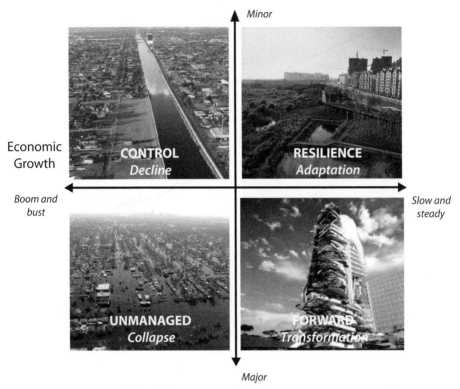

Minor

Economic
Growth

Boom and
bust

CONTROL
Decline

RESILIENCE
Adaptation

UNMANAGED
Collapse

FORWARD
Transformation

Slow and
steady

Major

FIGURE 5.12   Hypothetical scenarios resulting from two uncertain variables: climate change (major versus minor) and the economy (boom and bust versus slow and steady). Photographs (*clockwise from upper left*): 17th Street Canal, David Grunfeld Landov Media; Qunli National Urban Wetland, Turenscape; EDITT tower, Hamzah & Yeang; Hurricane Katrina, National Oceanic and Atmospheric Administration.

four alternative scenarios (figure 5.12). Each of these futures results from the interactions of uncertain variables, such as climate change and the economy. How will they affect carbon and nitrogen budgets or biodiversity over the long term?

Looking at figure 5.13, we can imagine the dramatic difference that could result from major climate changes under an unstable boom-and-bust economy compared to a slow and stable one. At the upper left, we see that limited pressure from climate change, coupled with an unstable economy, might cause people to be less willing, and less able, to implement

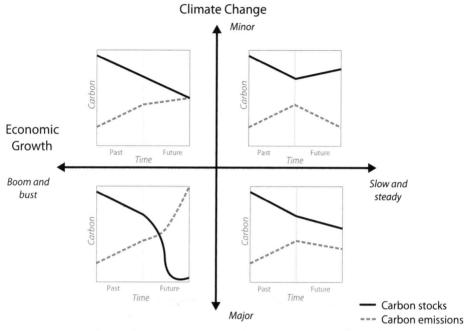

FIGURE 5.13 Hypothetical trends in the directions of slow and fast variables under plausible scenarios, showing substantial shifts from historical trends.

proactive conservation policies. Meanwhile, a stable economy, shown at the upper right, could lead to technological innovation. On the other hand, frequent perturbations from major climate impacts, coupled with economic instability, might lead to crisis and eventual collapse. Or, under a stable economy, they could instead lead to a completely different environment: one that encourages learning, innovation, and the emergence of new solutions.

Predictive models generate probabilities on the basis of observed dynamics. For example, we can estimate the probability distribution of carbon emissions given uncertain economic trajectories. But we cannot predict unexpected interactions among multiple uncertain driving forces, such as the possible trajectories that could result from interactions between major and minor climate change and a slow and steady versus an unstable boom-and-bust economy. Nor do we know the probability distributions of the impacts that could result from such interactions. The lower right quadrant of figure 5.14 shows that the actual impact could

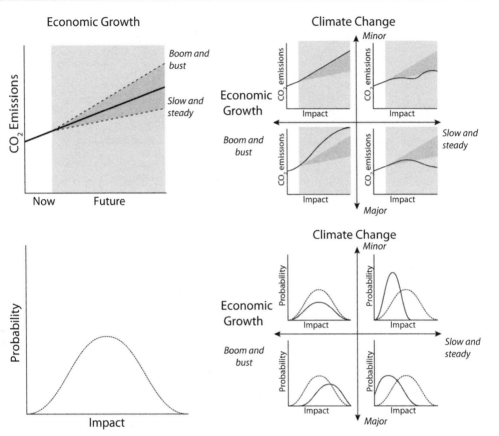

FIGURE 5.14   Probability distributions. The lower right graph shows that the actual impact of climate change could fall outside the predicted probability distribution of model projections (represented by the dotted lines).

very well fall outside of the predicted probability distributions of model projections (here represented by the dotted curves).

At the core of our challenge is the inevitable uncertainty of dynamic, coupled human-natural systems. Under plausible scenarios, trends in the directions of slow and fast variables could shift substantially from historical trends. For example, by assuming major climate changes and the boom-and-bust economy shown on the lower left, we can imagine rapid land conversion and an increase in logging; meanwhile, increasing energy

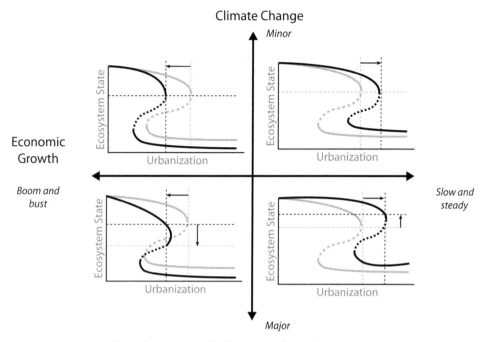

FIGURE 5.15 Hypothetical example illustrating how relationships between patterns of urban development and ecosystem function may vary under different scenarios.

demands and lack of regulatory power might lead to escalating emissions. Here the equilibrium curves associated with two alternative development patterns are represented under these plausible futures (figure 5.15). We might expect that the urban pattern would make dramatic differences for carbon stocks and fluxes and ecosystem resilience. But it might make no difference at all, depending on how these trajectories influence mechanisms and trade-offs among ecosystem functions across multiple scales.

Diversity and Modularity

I hypothesize that the diversity of patterns of urbanization maintained within a region and across regions might control the resilience of urban ecosystems (figure 5.16). Resilience depends on variable environmental and human conditions as well as on the unique history of each region. I suggest that the diversity of patterns might control resilience because

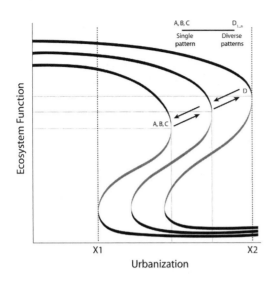

FIGURE 5.16   The diversity of patterns may control the resilience of urbanizing regions. A heterogeneous landscape composed of diverse patterns (D) is more resilient than any of the individual patterns.

this very diversity expands a region's capacity to adapt to a wider range of conditions and alternative futures. In the most recent examples of extreme climate events that tested the resilience of large urban regions across the world, the greatest surprises are in what worked (e.g., communication during Hurricane Sandy in 2013). Consider the countless ways in which unintended functions and flexibilities created by the diverse urban infrastructure (e.g., the transportation system) provided alternatives and ways out. An example is the vital role that bicycle infrastructure played during Sandy. With a flooded subway and shortage in gasoline, New Yorkers relied on the extended network of bike lanes, which provided them an alternative for commuting and allowed volunteers to bypass gridlocked traffic in their relief efforts. Most of all, they allowed communities to imagine ways to innovate and to adapt to rapidly changing conditions. Policies that aim at a single optimal pattern of urban development will eventually reduce the resilience of urbanizing regions in the face of uncertainty.

Implications for Planning: Emerging Principles

Planning agencies in urbanizing regions face unprecedented challenges: rapid environmental change places enormous pressure on their ability to support urban populations while maintaining a healthy ecosystem.

Agencies must make decisions about managing growth and investing in infrastructure while simultaneously providing human and ecosystem services. Strategic decisions about urban infrastructure and growth management are based on our assessment of the past and our expectations for the future. But complexity and uncertainty make the future increasingly unpredictable.

Climate change is expected to have significant impacts on essential human services (e.g., supplies of water and energy) and ecological functions (e.g., primary production) in urban areas. Potential regime shifts in coupled human-natural systems will inevitably surprise us. An inherent tension exists between resilience and transformation. How can planning help cities to enhance their adaptive capacities and thus facilitate transformation?

I propose five principles that we can use to rethink planning as a paradigm for resilience and for transforming hybrid ecosystems:

1. *Complexity*: Create and maintain diverse development patterns that support diverse human and ecosystem functions.
2. *Resilience*: Focus on maintaining self-organization and increasing the capacity to adapt instead of aiming to control change and to reduce uncertainty.
3. *Uncertainty*: Expand the ability to consider uncertainty and surprise by designing strategies that incorporate uncertainties and are robust to the most divergent plausible futures.
4. *Adaptation*: Create options for learning through experimentation, and opportunities to adapt through flexible policies and strategies that mimic the diversity of environmental and human communities.
5. *Transformation*: Expand capacity for change through transformative learning by challenging assumptions and actively reconfiguring problems.

6

Eco-Evolution on an Urban Planet

WHAT ROLE DO HUMANS PLAY IN THE EVOLUTION OF EARTH? CAN the emergence and rapid development of cities change the course of Earth's evolution? Can they determine the probability of crossing thresholds that will trigger abrupt change on a planetary scale? A great challenge for urban ecology in coming decades is to understand the role humans play in eco-evolutionary dynamics. If, as emerging evidence shows, rapid evolutionary change affects ecosystem functioning and stability, current rapid environmental change driven by urbanization might have significant implications for ecological and human well-being on a relatively short time scale. Humans are major drivers of microevolutionary change. At the same time, novel interactions between human and ecological processes may produce unprecedented expressions and opportunities for innovation. Understanding the mechanisms by which cities mediate eco-evolutionary feedback will provide important insights into how to maintain ecosystem function on an urbanizing planet.

Increasing evidence shows that cities are driving significant changes in wildlife, including plants, fungi, and other organisms. Songbirds are becoming tamer and bolder in response to the novel urban environments (Atwell et al. 2012) and are changing their tunes to keep urban background noise from masking their acoustic signals (Desrochers 2010; Dowling, Luther, and Marra 2011). Recent studies indicate that spiders are getting larger and even doing better in cities than in their natural habitats (Lowe, Wilder, and Hochuli 2014). Fish have adapted to cope with poisons in urban waters (Whitehead et al. 2010), while earthworms are well able to tolerate contaminants in soil (Kille et al. 2013). Seeds produced by weeds that grow in our parking lots do not travel very far compared to those growing in the countryside (Cheptou et al. 2008).

Humans are selective agents that determine which species can live in cities and cause organisms to undergo rapid evolutionary change. The urban habitat is not simply altering biodiversity by reducing the number and variety of native species. Urban-driven changes in biodiversity are more subtle and, to a certain extent, unexpected. Many organisms, including insects, birds, fish, mammals, and plants, are adapting to the new environment by changing their physiology, morphology, and behaviors. Microevolutionary change is taking place not only in tropical forests, but also in our backyards. The challenge for urban ecologists is to determine whether these changes might affect ecosystem function at the planetary scale. This question can be answered only by significant studies over the long term. Yet building on the emerging evidence, it is possible to articulate hypotheses linking urbanization to rapid evolution and to examine the role that urbanization could play in evolutionary dynamics.

Expanding the New Synthesis

Eco-evolutionary feedbacks—reciprocal interactions between ecological and evolutionary dynamics on contemporary time scales—were hypothesized over half a century ago (Pimentel 1961), but only recently have they been tested empirically (Schoener 2011). There is significant evidence that changes in ecological conditions drive evolutionary change in species traits that, in turn, alters ecological interactions (Endler 1986; Odling-Smee, Laland, and Feldman 2003). Yet despite the remarkable progress in studying eco-evolutionary feedbacks over the past decade, empirical studies are still limited, and potential implications for environmental change and the evolution of species are only beginning to emerge (Odling-Smee, Laland, and Feldman 2003; Post and Palkovacs 2009; Stockwell 2003). In particular, we do not know what role human activity plays in reciprocal interactions between ecological and evolutionary processes.

Earlier assumptions about the different time scales of ecological and evolutionary processes have shaped the unidirectional character of most empirical eco-evolutionary studies and can partly explain our lack of curiosity about the human role in shaping the evolutionary trajectory of Planet Earth. But recent evidence that significant evolutionary change does occur on a short time scale urgently challenges both ecologists and evolutionary biologists to redefine the dynamic interplay between the two fields and to understand interactions between human agency and eco-evolutionary feedback across different levels of biological organization.

Humans are major drivers of microevolutionary change (Hendry, Farrugia, and Kinnison 2008; Palkovacs et al. 2012). In human-dominated environments, selection pressures acting on traits can affect population dynamics by changing organisms' rates of survival or reproductive success, leaving a genetic signature that might affect community dynamics and ecosystem functions (J. N. Thompson 1998). Phenotypic trait changes resulting from changes in gene frequencies might affect population dynamics through changes in demographic rates (Pelletier, Garant, and Hendry 2009). Genetic signatures have been observed in the population dynamics of several organisms, including birds, fish, arthropods, rodents, land plants, and algae (Hendry, Farrugia, and Kinnison 2008; Palkovacs and Hendry 2010). Effects at the community level might result from predator-prey interactions, parasite-host relationships, mutualism, and competition (Fussmann, Loreau, and Abrams 2007). These effects drive changes in energy and material fluxes that in turn influence ecosystem functions, such as primary productivity, nutrient cycling, hydrological function, and biodiversity (Matthews et al. 2011), which provide essential services for human well-being (J. N. Thompson 1998).

The emergence and rapid development of cities across the globe might represent a turning point in human-driven eco-evolutionary dynamics in ways we do not yet completely understand. In cities, subtle eco-evolutionary changes are at play—and they are occurring at an unprecedented pace. Urbanization mediates eco-evolutionary feedback by simultaneously changing habitat and biotic interactions and driving socioeconomic transitions toward an increased pace of life. The extraordinary concentration of people and activities in cities provides major opportunities to achieve economies of scale, but it also intensifies the use of energy and its environmental impacts. While cities accelerate the transition to efficient technologies, technological innovation provides access to resources from distant regions, promoting positive feedbacks (Bettencourt and West 2010). Cities are not simply altering biodiversity by reducing the number and variety of native species. Humans are selective agents that determine which species can live in cities and cause organisms to undergo rapid evolutionary change. Many organisms, including arthropods, birds, fish, mammals, and plants, are adapting to the new environment by changing their physiology, morphology, and behaviors.

During the past decade, increasing evidence has suggested not only that species diversity matters to ecosystem function, but also that what determines the magnitude of its effect is species identity (Cardinale, Hillebrand, and Charles 2006). By focusing on functional groups, ecological scholars have recently begun to investigate the extent to which functional substitutions alter a variety of properties, such as primary productivity, decomposition rates, and nutrient cycling (Loreau 2010), as well as ecosystem stability and resilience (Loreau and de Mazancourt 2013). Biodiversity might provide "insurance," a buffer that can maintain ecosystem function in the presence of environmental variability, since different species respond differently to environmental fluctuations (Loreau and de Mazancourt 2013). Recent studies indicate that "response diversity"—the variability in responses of species within functional groups—is what sustains ecosystems in the context of rapid environmental change (Mori, Furukawa, and Sasaki 2013).

Humans can affect the composition and functional roles of the species in ecosystems both directly, by reducing the overall number of species, and selectively, by determining phenotypic trait diversity (Partecke 2014). Individual species can control processes at both the community and ecosystem levels (Lawton 1994); thus diversity could have a strong effect on such processes, as changes in diversity affect the probability that given species will occur among potential colonists (Tilman 1999).

By bringing human agency into the study of eco-evolutionary feedback, we can begin to articulate and test a series of hypotheses about key mechanisms linking biodiversity and ecosystem function (Pickett et al. 2011) and about potential feedbacks between evolution and ecosystem dynamics on a human-dominated planet (Matthews et al. 2011; Palkovacs et al. 2012). But to fully appreciate the implications of including humans in such a framework, we must consider several levels of human interactions with ecological and evolutionary processes.

In the following sections, I first review examples of human-driven eco-evolutionary feedbacks to articulate emerging hypotheses on how urbanization might drive eco-evolutionary dynamics and influence planetary change. Then, by focusing on documented signatures of trait change, I identify emerging mechanisms that link urbanization to both eco-evolutionary dynamics and potential feedbacks on ecosystem function. I next elaborate on how co-evolutionary interactions between species or genes can mediate evolutionary feedbacks through either strictly genetic co-evolution or through gene-culture co-evolution. Finally, I discuss how the rapid change associated with urbanization can give rise to a range of feedbacks that govern the behavior of evolutionary change, and I consider their potential implications for either promoting or buffering potential regime shifts.

Integrating Humans into Eco-Evolutionary Dynamics

Increasing evidence shows that humans influence evolutionary processes by changing speciation and extinction patterns (Palumbi 2001). Humans are creating and dispersing thousands of synthetic compounds and thereby altering bacteria, insects, and other organisms. By hunting, fishing selectively, and reconfiguring the planet's surface, humans have triggered a wave of extinction comparable to the five mass extinctions in Earth's earlier history (Dirzo et al. 2014). The anthropogenic signatures of planet-scale changes are most evident in urbanizing regions. Increasing evidence shows that as humans interact with niche construction through urbanization, they alter the structure and function of communities and ecosystems (Donihue and Lambert 2015; Marzluff 2012). Yet the evolutionary consequences of urbanization and the mechanisms by which dense human settlements affect selective processes are not well known.

Eco-evolutionary biologists have developed several expressions to formalize eco-evolutionary feedback. I revise the general definition from Post and Palkovacs (2009) by explicitly identifying urbanization as a variable that intervenes in the interplay between ecological and evolutionary dynamics (Alberti et al. 2003). By applying the Palkovacs and Hendry (2010) framework, we can start to identify urban-driven changes in the attributes of populations, communities, and ecosystems that influence phenotypes via selection and plasticity and potential feedbacks (figure 6.1). Examples of eco-evolutionary feedbacks associated with urbanization (both hypothesized and documented) have been illustrated for many species of birds, fish, plants, mammals, and invertebrates (table 6.1).

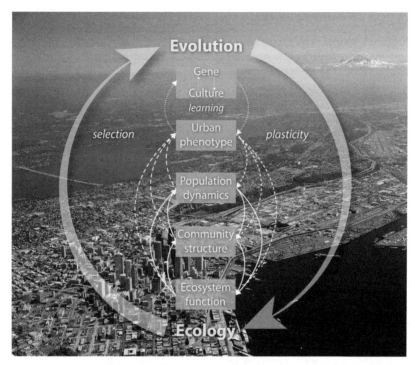

FIGURE 6.1   Eco-evolutionary dynamics in urbanizing ecosystems. Changes in
gene frequencies might translate into phenotypic trait changes (e.g., physiology,
morphology, behavior) that affect demographic rates (e.g., reproduction, survival,
or dispersal) and ultimately population dynamics (e.g., numbers of individuals
and population persistence), community structure (e.g., species richness or diversity), and ecosystem function (e.g., nutrient cycling, decomposition, and primary
productivity) (dashed lines). These changes can cascade among levels of ecological
organization (continuous lines) and ultimately affect the trajectory of evolution
(loops represented by dotted lines). To fully understand the potential implications
of eco-evolutionary feedbacks in urbanizing environments, it is critical to expand
the framework of eco-evolutionary feedback to include genetic and cultural co-
evolution through both inheritance and social learning. The general framework
for eco-evolutionary feedback is adapted from Palkovacs and Hendry 2010. Background photograph reproduced with permission from Aerolist-photo.com.

Building on examples documented in the literature (Palkovacs et al.
2012; Palkovacs and Hendry 2010; Post and Palkovacs 2009; Schoener
2011), I articulate four overarching hypotheses (H1–4) linking urbanization to eco-evolutionary dynamics and its potential role in promoting or
buffering eco-evolutionary change:

Table 6.1. *Examples of urban signatures: A synoptic table of documented examples of cases of trait change in fish, invertebrates, birds, mammals, and plants; their human drivers; and their ecological effects*

Species	Driver	Traits	Mechanism	Ecological Effects	References
		Fish			
Pacific salmon (*Oncorhynchus* spp.)	Dam construction	Body shape	Phenotypic Genetic	Trophic interactions*	Franssen, Stewart, and Schaefer 2013; Haas, Blum, and Heins 2010
Alewife (*Alosa pseudoharengus*)	Dam construction	Gape size Migratory behavior	Phenotypic	Trophic cascade Nutrient subsidies	Post and Palkovacs 2009; Walters, Barnes, and Post 2009
Killfish (*Fundulus heteroclitus*)	PCB contamination	Tolerance to toxicity	Genetic	Trophic cascade	Whitehead et al. 2010
Largemouth bass (*Micropterus salmoides*)	Recreational fishing	Growth rate Vulnerability to angling	Genetic	Social behavior Trophic interactions*	Cooke et al. 2007; Philipp et al. 2009
Top predators (e.g., Atlantic cod, *Gadus morhua*)	Commercial fishing	Body size Metabolic rate	Phenotypic	Trophic cascade	Shackell et al. 2010
		Invertebrates			
Peppered moth (*Biston betularia*)	Industrial pollution Predation	Melanism	Genetic	Biodiversity	Kettlewell 1958

(continued)

Table 6.1. (continued)

Species	Driver	Traits	Mechanism	Ecological Effects	References
Daphnia (D. pulex)	Eutrophication	Resistance to toxins Cyanobacteria	Phenotypic	Trophic cascades*	Hairston et al. 2001
Daphnia (D. pulex)	Hydrological impact on predator-prey interaction	Reproduction	Phenotypic Genetic	Consumer dynamic*	Miner et al. 2012
Earthworm (Lumbricus rubellus)	Soil contamination; trace elements (e.g., arsenic)	Tolerance to metals	Phenotypic Genetic	Nutrient cycling*	Kille et al. 2013
Birds					
Dark-eyed junco (Junco hyemalis)	Heat island	Tail feathers	Genetic	Biodiversity Seed dispersal Biotic control	Yeh and Price 2004
Songbirds	Fragmentation	Wing shape	Phenotypic	Metapopulation dynamics*	Desrochers 2010
European blackcap (Sylvia atricapilla)	Supplemental feeding	Wing shape Beak shape	Phenotypic	Niche diversification*	Rolshausen et al. 2009
Great tit (Parus major)	Noise	Song acoustic-adaptation	Genetic	Biodiversity	Slabbekoorn 2013
European blackbird (Turdus merula)	Artificial light	Stress response behavior	Genetic	Metapopulation dynamics*	Partecke 2014

	Mammals			
Red fox (*Vulpes vulpes*)	New food	Body size	Genetic	Trophic Dynamic	Wandeler et al. 2003
White-footed mouse (*Peromyscus leucopus*)	Fragmentation	Protein coding Immune system	Genetic	Population growth Diseases	Harris et al. 2013
	Plants			
Plants	Elevated CO_2 concentration	Leaf nitrogen composition	Phenotypic	Consumer dynamics*	Hairston et al. 2001
Populus	Habitat modification	Leaf tannin levels	Genetic	Nutrient cycling	Whitham et al. 2006
Weed (*Crepis sancta*)	Fragmentation	Dispersal	Genetic	Metapopulation dynamics*	Cheptou et al. 2008

*Hypothesized.

H1: Genetic signatures of urban eco-evolutionary feedback can be detected across multiple taxa and ecosystem functions.

H2: Through urbanization, humans mediate interactions and feedbacks between evolution and ecology in subtle ways by introducing changes in habitat, biotic interactions, heterogeneity, novel disturbance, and social interactions.

H3: Humans affect eco-evolutionary feedback through both genetic and cultural changes resulting from their *co-evolutionary dynamics* with other social organisms.

H4: The hybrid nature of urban ecosystems—resulting from co-evolving human and natural systems—is a source of *innovation* in eco-evolutionary processes.

Detecting the Genetic Signatures of Urban Evolutionary Change

Urbanization alters natural habitats, leading to new *selection pressures* and *phenotypic plasticity* (figure 6.2). The significant decrease in biodiversity in cities is only the most apparent of several subtle changes associated with urbanization that have the potential to affect genetic diversity (Marzluff 2012; Partecke 2014; Pickett et al. 2011). Habitat modification and fragmentation might lead to genetic differentiation. New predators and competitors affect species interactions. Exposure to new pathogens in cities alters host-pathogen interactions and can influence species fitness via the immune system. Diverse forms of pollution, from toxins to noise and light, can favor species that adapt more efficiently than others to these new conditions. Furthermore, the socioeconomic transitions and technological shifts associated with urbanization significantly affect the scale and pace of ecological change, leading to rapid evolution beyond cities' boundaries.

New selection regimes have significant consequences for microevolutionary changes. At the same time, extreme turnover in biological communities might prevent urban populations from differentiating genetically and might impede evolutionary responses to the novel selective forces associated with urbanization (Shochat et al. 2006). Urbanization also induces phenotypic responses via phenotypic plasticity, but such responses might require genetic adaptation (Kawecki and Ebert

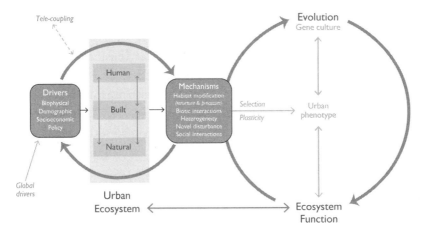

FIGURE 6.2   A conceptual framework of urban eco-evolutionary feedback. This is an integrated model to identify key mechanisms linking urban ecosystem dynamics to eco-evolutionary feedback. Key human drivers of change (in, e.g., climate, demographics, economics, and policy) influence eco-evolutionary dynamics through interactions among the human, natural, and built system components of the urban ecosystem. Highlighted are the emerging mechanisms through which urbanization drives eco-evolutionary dynamics: habitat change (structure and processes), biotic interactions, heterogeneity, novel disturbances, and social interactions.

2004; Partecke 2014). Heritable differences can accumulate through genetic accommodation, and plastic responses can generate new selection pressures (Richter-Boix et al. 2013). Complex genetic and cultural co-evolution among species and between humans and other organisms also play important roles through social learning (Marzluff 2012).

As shown in table 6.1, several studies have documented rapid human-driven trait changes (Hendry, Farrugia, and Kinnison 2008; Palkovacs et al. 2012), but only a few specifically examine the role of urbanization (box 6.1). Hendry et al. (ibid.) indicated that human-driven trait changes occur roughly twice as fast as those driven by nonanthropogenic forces. Recent reviews of adaptive evolution in urban ecosystems have begun to document studies of specific organisms (Marzluff 2012) and to synthesize a spectrum of applied techniques (Donihue and Lambert 2014). Examples of urban-driven eco-evolutionary feedbacks (both hypothesized and documented) have been illustrated for many species of birds (Desrochers 2010; Rolshausen et al. 2009; Yeh and Price 2004), fish (Carlson, Quinn, and Hendry 2011; Franssen, Stewart, and Schaefer 2013; Haas,

Blum, and Heins 2010; Williams et al. 2008), rodents (Harris et al. 2013), plants (Cheptou et al. 2008; Jacquemyn et al. 2012; Riba et al. 2009), and amphibians, as well as for diverse invertebrates (Post and Palkovacs 2009). A few notable examples (e.g., *Daphnia*) are revealing the reciprocal causal mechanisms that drive interactions between organisms and their environments, as new selective pressures can alter the population dynamics of multiple prey species, reconfigure trophic interactions, and ultimately affect multiple ecosystem functions (Matthews et al. 2011). Understanding the mechanisms by which species successfully adapt to human-driven changes and urban environments is critical to anticipating the future evolutionary trajectories of the urbanizing planet.

Box 6.1. Examples of Urban Eco-Evolution

One of the earliest examples of urban adaptation is the dark-eyed junco (*Junco hyemalis*), a genus of small, grayish American sparrows. In San Diego, California, for example, it has adapted its morphology and behavior: there these birds have shorter wings and tails compared to those that live in mountains, and they also have about 22 percent less white in their outer tail feathers (Yeh and Price 2004). Furthermore, Atwell et al. (2012) compared the urban bird to a population in forested areas and found that the urban ones have both bolder behavior and lower levels of corticosterone under stressful conditions.

Several scholars have documented evidence of changes in the morphological attributes of fish in response to the construction of dams and habitat changes (Carlson, Quinn, and Hendry 2011; Franssen, Stewart, and Schaefer 2013; Haas, Blum, and Heins 2010; Williams et al. 2008). The evolutionary changes of a keystone aquatic herbivore, *Daphnia*, provide evidence that selection pressures may have an impact on trophic interactions (Miner et al. 2012). Selection pressures that influence the composition of zooplankton communities result from the significant changes in biogeochemical cycles and physical templates that occur in urbanizing catchment areas, leading to rapid evolutionary change.

Evolutionary changes in the reproductive traits of plants in urban habitats are starting to be documented, particularly the effect of selective pressure on seed dispersal. Cheptou et al. (2008) found that the weed *Crepis sancta* disperses a significantly lower proportion of its seeds in urban patches than in unfragmented surroundings. Jacquemyn et al. (2012) also found that urbanization causes rapid evolution of seed sizes and seed dispersal.

Harris et al. (2013) studied white-footed mice (*Peromyscus leucopus*) in the New York metropolitan area and found a signature of directional selection in divergent urban ecosystems. These mice are the critical hosts for black-legged ticks, which carry and spread the bacterium that causes Lyme disease. Superabundant mouse populations allow more ticks to survive and lead to predictable spikes in human exposure to Lyme.

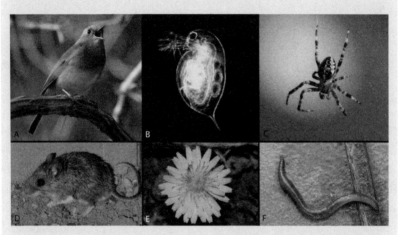

FIGURE B6.1   *Top row, left to right*: In San Diego, California, the dark-eyed junco (*Junco hyemalis*) has adapted its morphology and behavior to the novel urban environment. The water flea *Daphnia* plays a pivotal role in the functioning of pelagic freshwater food webs (Miner et al. 2012). The orb-weaving spider (*Nephila plumipes*) is growing larger in urban areas (Lowe, Wilder, and Hochuli 2014). *Bottom row, left to right*: The white-footed mouse (*Peromyscus leucopus*), a common resident of New York City's forest fragments, exhibits signatures of directional selection in urban ecosystems (Harris et al. 2013). *Crepis sancta*'s seed dispersal has evolved rapidly (Cheptou et al. 2008). Earthworms (*Lumbricus rubellus*) have adapted genetically to a series of soil contaminants (Kille et al. 2013). Photographs: songbird: Paul Albertella; plankton: Paul Heber; spider: olavgg; mouse: J. N. Smart; flower: Bernard Dupont; earthworm: Belteguese.

Another "urban exploiter" is the spider. Sydney, Lowe, Wilder, and Hochuli (2014) found that in urban environments, the orb-weaving spider (*Nephila plumipes*), indigenous to the Australian countryside, grows larger and exhibits higher fecundity than its counterparts in the countryside. They collected and measured more than two hundred of these spiders from urban, semiurban, and semirural sites in and around Sydney.

Earthworms (*Lumbricus rubellus*) have adapted genetically to a series of soil contaminants. In a recent study, two teams in the United Kingdom led by Peter Kille (Kille et al. 2013) at Cardiff University, used a new eight-thousand-element microarray to describe the transcriptome profile of *L. rubellus* exposed to copper, cadmium, the polyaromatic hydrocarbon fluoroanthene, and the agrochemical atrazine. They revealed that the toxic chemicals induced subtle changes in patterns of earthworm gene expression that let the worms evolve adaptive mechanisms to deal with soil pollution. •

Emerging Urban Eco-Evolutionary Mechanisms

Urbanization mediates eco-evolutionary feedbacks through several filters that operate simultaneously across multiple scales. Urban development changes habitat structures and biogeochemical cycles, modifies disturbance regimes, and introduces species (e.g., hosts, pathogens, and predators), creating novel habitats (figure 6.2, earlier). Urban environments can facilitate speciation by bringing together previously isolated species or by isolating populations through habitat transformation (Partecke 2014). Changes in habitat and selective forces increase the chances of extinction (Marzluff 2012). In addition to changes in the physical template, humans in cities modify the availability of resources and their variability over time, buffering the effects of resource variability on community structure (Shochat et al. 2006). Although each filter can be identified as an independent driver, their consequences cannot be understood in isolation. Furthermore, because cities increase the pace of life (Bettencourt and West 2010) and amplify tele-coupled interactions, human activities also have impacts on distant places (Liu et al. 2013).

Mechanisms by Which Urbanization Affects Evolutionary Dynamics

Hypotheses about how humans in urbanizing environments impact eco-evolutionary feedbacks can be articulated around key mechanisms that influence species diversity, ecosystem function, and, ultimately, human well-being (Partecke 2014; Shochat et al. 2006). In urban environments, selective changes are caused by the combined effects of changes in *habitat structure* (e.g., loss of forest cover and connectivity) and processes (e.g., biogeochemical and nutrient cycling) and changes in *biotic interactions* (e.g., predation). Humans in cities also mediate eco-evolutionary interactions by introducing *novel disturbances* and altering habitat *heterogeneity* (McKinney 2006) (box 6.2). These mechanisms vary across the urban gradient (box 6.3). Complex interactions resulting from changes in habitat and biotic interactions, coupled with the emerging spatial and temporal patterns of resource availability, might produce new *trophic* dynamics (e.g., shifts in control from top-down to bottom-up) (Shochat et al. 2006).

Box 6.2. The Homogenization Hypothesis

The effects that human actions have on spatial heterogeneity in urbanizing regions is well documented, but we know less about how heterogeneity varies with scale, partly because studies have tended to focus primarily on aggregated measures (Pickett et al. 2011). At the scale of meters or smaller, urbanization might reduce the heterogeneity of land cover, but at the patch level, it might introduce highly heterogeneous new biophysical conditions as the varied behaviors of landowners result in fragmented management patterns. As the scale increases, a further reduction in heterogeneity could occur due to consistent patterns of urban development and habitat fragmentation. McKinney (2006) advanced the hypothesis that urbanization causes global homogenization. As cities expand, urban regions are maintained in a state of disequilibrium from the local natural environment, so that habitats across urban sites are more similar to one another than to their respective adjacent natural environments (Groffman et al. 2014). Urban management and the built infrastructure can artificially reduce the variation—in both space and time—of resource availability, thus altering seasonal variations and dampening temporal variability (Shochat et al. 2006). Some species thrive when they have less variation to endure, and thus their urban populations rise. A well-known example is the grey-headed flying fox (*Pteropus poliocephalus*), a large nomadic bat from eastern Australia, which became established in Melbourne when a heat island effect led to long-term climatic changes. Parris and Hazell (2005) found that human activities have increased temperatures and effective precipitation in central Melbourne, creating a more suitable climate for the flying fox to camp. Changed habitat, with more water and food available, interacts with biotic, trophic, and genetic processes and might help some species adapt to urban environments.

Changes in temporal variability in urban ecosystems are driven by both human structures and high inputs of resources. A good example of such a change is the buffering effect that microclimatic changes associated with urbanization can have on habitat: in temperate cities, heat islands can extend the growing season, but in desert cities, they can cause extended droughts. Shochat et al. (2006) reported that the heat island in Phoenix, Arizona, has increased the stress on cotton plants (*Gossypium hirsutum*). Highly managed green areas in temperate cities such as Seattle, Washington, and irrigation in semiarid cities such as Salt Lake City, Utah, provide water for plants throughout the year, with subtle effects on wildlife. ●

In urban ecosystems, however, changes in ecological dynamics are only part of the picture. What makes urban ecosystems unique is the presence of people. Cities are shaped by social interactions (Bettencourt 2013) and cultural evolution (Marzluff 2012), with significant consequences for co-evolution between humans and other species and for the pace of change (Bettencourt and West 2010). One such consequence is the phenomenon of tele-coupling: emerging interactions between distant natural and human systems (Liu et al. 2013) that expand the edge of urban-driven eco-evolutionary change beyond the city itself and accelerate its dynamics.

Box 6.3. Urban Ecological Gradients

Patterns of development produce different landscape signatures—spatial and temporal changes in ecosystem processes that, in turn, influence biodiversity (Alberti 2008)—that vary along an urban-to-rural gradient, ranging from the urban core to suburban and exurban areas to rural areas and intact forest. Emerging hypotheses on mechanisms linking urbanization patterns to changes in species traits are based on the evidence that patterns of urbanization affect natural habitat and biotic interactions across the urban-to-rural gradient in subtle ways (Pickett, Wu, and Cadenasso 1999). Hypothesized urban landscape signatures are represented in figure B6.2 in relationship to each mechanism that initial evidence suggests may vary along a hypothetical urban gradient (*x*-axis) (Pickett et al. 2011).

Ecosystem functions along a gradient of urbanization are simultaneously affected by changes in habitats and biotic interactions (figure B6.2). Rates of forest conversion and loss of native habitat are highest at the urban fringe, as forest conversion has already occurred at the urban core. Forest connectivity declines closer to the urban core. Resource availability is kept artificially high at the urban fringe because of human inputs, but the variability of available resources declines. Disturbances increase with urbanization.

We can expect a steady decline in native species and an increase in extinction toward the urban core; meanwhile, colonization by early successional and synanthropic species peaks at the urban fringe and then declines at the urban core (Marzluff 2005). Predation declines toward the urban core, although not steadily, due to predation by urban pets. Parasitism is higher closer to the urban core. Humans might facilitate parasites in the suburbs by, for example, putting out bird nest boxes. Insect species that are facilitated in nature by woodpeckers are facilitated in suburbs by woodpeckers, people, and other birds; for example, chickadees and swallows nest in nest boxes and lights. We can expect that novel competitions will emerge as landscapes urbanize (Marzluff 2012).

Together, habitat modification and changes in biotic interactions

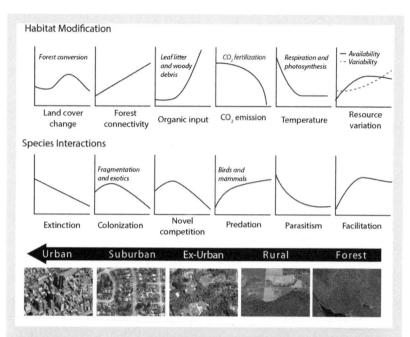

FIGURE B6.2   Hypothesized urban landscape signatures emerging from the literature: the relationships of key mechanisms associated with habitat modification and species interactions represented on the y-axes, and a hypothetical urban-natural gradient. The hypothesized signatures highlight the complexity of the relationships and do not apply across all urban-natural gradients (for example, in regions where adjacent wild lands are steppe, savanna, or desert). Aerial photographs: Google Maps 2015.

lead to evolutionary responses in species and ecosystem functions. In cities, humans modify the mechanisms that control the spatial and temporal variability of nutrient sources and sinks (Alberti 2008; Kaye et al. 2006).

In New York, urban forests exhibit faster rates of litter decomposition and nitrification than rural forests (Pickett et al. 1999). The heat island effect and the introduction and colonization of nonnative earthworms in urban forests were hypothesized to drive these results. However, empirical studies of the underlying processes and mechanisms linking urbanization patterns and ecosystem dynamics are still extremely limited (Alberti 2008). ●

Habitat Modification

Land-cover conversion and rapid loss of native habitat are major drivers of microevolutionary change. Habitat patches and their species communities are often isolated from each other by a matrix of built environments. Fragmentation of natural patches is one of the best-known impacts of human activities on the diversity, structure, and distribution of vegetation as well as on the movement of resources and organisms among natural patches. New barriers make dispersal difficult and potentially penalize less mobile organisms (Rebele 1994). Furthermore, changes in productivity—the rate at which energy flows through an ecosystem—might explain patterns of species diversity along the urban-to-rural gradient (Waide et al. 1999), although studies have produced contradictory results. Net primary production mediates the relationship between anthropogenic land-cover change and the richness of both faunal and plant species, but the relationship varies with taxa and scale and across biomes (Mittelbach et al. 2001). Human activities drive direct and indirect changes in the distribution of resources, which can peak at the urban fringe, simultaneously reducing their variability.

Biotic Interactions

Urban development creates new opportunities and challenges for species competition and predation, both as exotic species are introduced and as invasive species migrate in, taking advantage of poorly integrated communities and patches in the urban setting. This sometimes results in a colonization process, as more frequent introductions of exotic species translate into invasions (Faeth et al. 2005). Examples of this phenomenon abound (Marzluff 2012). Along a 140-kilometer urban-to-rural environmental gradient originating in New York City, McDonnell et al. (1997) found lower levels of both earthworm biomass and abundance in urban forests compared to rural ones, likely because of introduced species. Urbanization also alters the way species are distributed and interact with one another (Hansen et al. 2005). Marzluff (2005) developed a series of testable hypotheses about how urbanization affects colonization and extinction in determining local diversity; he found that while diversity still emerges as the balance between extinction and colonization, species invasion plays a prominent role.

Heterogeneity

Cities are "human habitats" and are designed and managed to best support human functions, and thus we find lower diversity of microclimates and species among urban sites as compared to adjacent, natural ecosystems (box 6.2). The diversity of species in urbanizing regions is greatly affected by the quality of habitat and the template of resources. Habitat *heterogeneity* allows for greater niche differentiation and, hence, more species. Humans in cities affect habitat quality and resource availability by changing their heterogeneity in space and time. An example of change in temporal heterogeneity is the buffering effect created by microclimatic changes associated with urbanization. Heat islands can extend the growing season in temperate cities, but they extend droughts in desert urban areas (Shochat et al. 2006). Despite the high numbers of small patches with different environmental conditions (Kowarik 2011), habitat changes associated with urban land uses act as filters in urban species composition, with clear winners and losers, and losses of native species drive the homogenization of ecological structures and functions (Groffman et al. 2014). Yet cities worldwide still retain native species (Aronson et al. 2014).

Novel Disturbance

Ecosystem disturbances affect the availability of resources (e.g., water and nutrients), ecosystem productivity, and species diversity (Pickett, Wu, and Cadenasso 1999). Urbanization modifies existing disturbance regimes through, for example, fire and flood management, and creates novel disturbances, such as new or disrupted dispersal pathways or introduced species. Human-induced disturbances in urban environments maintain urban habitats at an early successional stage (McKinney 2006; Pickett, Wu, and Cadenasso 1999). Furthermore, the patchy distribution of urban habitats, combined with their varying degree of human-induced disturbances, results in a number of succession paths across habitat patches (Kowarik 2011). Cardinale, Hillebrand, and Charles (2006) suggested that disturbance can moderate relationships between biodiversity and ecosystem functioning in two ways: it can increase the chance that diversity generates unique system properties and it can suppress the probability of ecological processes being controlled by a single taxon.

Social Interactions

Urbanization also changes the dynamics of social interactions among people (Bettencourt 2013) and between people and other species (Marzluff 2012). Perhaps the most significant quality that distinguishes cities from other systems is their pace of change. By examining a large set of data on diverse aspects that characterize urban regions, Bettencourt and West (2010) observed that while cities exhibit scaling relationships similar to those that biologists have found for organisms' molecular, physiological, ecological, and life-history attributes, some relationships have no analog in natural systems. In nature, the networks and interactions that sustain biological organisms and ecosystems are dominated by economies of scale, or sublinear scaling. In cities, environmental changes are driven by social interactions that operate in exactly the opposite fashion, showing superlinear scaling. The larger the city, the faster its pace of life (Bettencourt 2013).

Tele-Coupling

Cities are networked far beyond their own physical edges. Their functions depend on highly interconnected infrastructures and on flows of material, energy, and information from both proximate regions (e.g., via hydroelectric dams) and distant ones (e.g., via trade and telecommunication). Distant coupled human-natural interactions are more prevalent and occur at higher speeds (Liu et al. 2013). Such complex interactions in tele-coupled systems make it particularly challenging to understand the potential eco-evolutionary implications and the feedbacks associated with urbanization and to anticipate potential crossings of thresholds or the outcomes of such crossings. On a human-dominated planet, tele-coupling challenges the ecosystem concept implied in eco-evolutionary studies to expand both its boundaries and its relationships to include distant interactions and new potential cross-scale feedbacks (Liu et al. 2013).

Mechanisms by Which Evolutionary Change Affects Urban Ecosystems

Researchers are increasingly documenting how phenotypic evolution might affect ecosystem functions (Loreau 2010). Individual trait variation has significant implications for ecosystems' productivity and their stability; thus, according to Matthews et al. (2011), it represents a natural intersection between evolutionary biology and ecosystem science.

Linking Phenotypic Evolution to Urban Population and Community Dynamics

Table 6.2 identifies examples of potentially heritable traits that might directly or indirectly affect ecosystem functions and for which there is evidence of evolutionary response to environmental changes driven by urbanization (Matthews et al. 2011; Palkovacs and Hendry 2010; Shochat et al. 2006). The evolution of organisms' traits that control ecosystem processes could lead to significant changes in ecosystem functions through organisms' ability to alter their environment and their selective regimes (Odling-Smee, Laland, and Feldman 2003). For example, primary productivity is associated with consumer traits that regulate consumers' demand for resources. Evolution in such traits can affect nutrient cycling and, ultimately, the magnitude and spatial distribution of primary production (McIntyre et al. 2008). Seed dispersers have a significant impact on plant diversity and their functional role in urban ecosystems. A great diversity of organisms modify the physical structures of estuarine and coastal environments, particularly dune and marsh plants, mangroves, sea grasses, kelps, and infauna. Evolution in traits underlying organisms' ecosystem-engineering effects has potentially significant functional impacts.

Matthews et al. (2011) examined how the evolution of ecosystem-effect traits can directly or indirectly affect ecosystem functions by influencing ecosystem processes via environmental, population, and community dynamics. For example, the evolution of plants' photosynthetic traits could, in turn, alter rates of primary production and carbon sequestration in terrestrial ecosystems. CO_2 concentrations can also affect the growth rate and phenotype of algae, which could alter rates

Table 6.2. Mapping urban-driven heritable trait change to ecosystem function: Documented changes in heritable traits, urban drivers, and hypothesized mechanisms of eco-evolutionary feedback

Urban Habitat		Heritable Trait	Eco-Evolutionary Feedback		References
Habitat Modification	Biotic Interactions	Physiological	Ecosystem Function	Feedback Mechanism	
CO_2 concentration		Photosynthetic rate (algae)	Primary productivity Nutrient cycling	CO_2 effect on algae growth rate and phenotype	Collins and Bell 2004
CO_2 concentration	Food web and trophic interaction	Leaf nitrogen composition (plants)	Nutrient cycling	Herbivore density and feeding behavior	Stiling et al. 2003
Toxic chemicals	Trophic interaction	Endocrine system/hormones (fish)	Nutrient cycling Biodiversity	Impaired signaling pathways	Shenoy and Crowley 2011; Walters, Barnes, and Post 2009
Toxic chemicals Noise Light	Trophic interaction	Endocrine system/hormones (birds)	Biotic control Seed dispersal Biodiversity	Impaired reproductive and immune functions	Shenoy and Crowley 2011
Metals		Tolerance to metals (earthworms)	Nutrient cycling Decomposition	Increased numbers and biomass of earthworms	Kille et al. 2013
Nutrient loads Eutrophication	Trophic cascade	Resistance to toxic cyanobacteria (Daphnia)	Primary productivity Nutrient cycling	Consumer-resource dynamics	Hairston et al. 2001

(continued)

Table 6.2. (continued)

Urban Habitat		Heritable Trait	Eco-Evolutionary Feedback		References
Habitat Modification	Biotic Interactions	Physiological	Ecosystem Function	Feedback Mechanism	
Morphological					
Hydrological connectivity	Trophic interactions Predator-prey interaction	Body shape/size (fish)	Biodiversity Nutrient cycling	Effects on life history of zooplankton (*Daphnia*)	Walsh et al. 2012
Emissions Heat		Plumage (birds)	Biodiversity	Colonization	Yeh and Price 2004
Forest fragmentation Food	Novel competition	Wing shape (birds)	Biotic control Seed dispersal Biodiversity	Niche diversification Metapopulation dynamic	Desrochers 2010; Rolshausen et al. 2009
Behavioral					
Hydrological connectivity	Trophic interactions	Migratory propensity (fish)	Nutrient cycling Biodiversity	Relative energetic or survival costs of migration	Post et al. 2008
Forest fragmentation	Competition for food Predation risk	Foraging (birds)	Biodiversity	Efficiency in exploiting food resources	Shochat et al. 2006

Artificial lighting Noise	Competition for territories	Syndromes (neophilic and neophobic) (birds)	Biodiversity	Colonization	Miranda et al. 2013
Heat island Food	Predation risk	Migratory propensity (birds)	Biodiversity Biotic control	Meta-population dynamic	Partecke 2014
Phenological/Life History					
Heat island Artificial lighting Food	Breeding density	Time and duration of reproduction (birds)	Biodiversity	Colonization	Partecke 2014
Hydrological connectivity	Predator-prey interaction	Time and reproductive effort (*Daphnia*)	Primary productivity Nutrient cycling	Consumer dynamic	Walsh et al. 2012
Fragmentation		Dispersal (seeds)	Biodiversity Nutrient cycling	Metapopulation dynamics	Cheptou et al. 2008

of primary production and carbon sequestration in aquatic ecosystems (Collins and Bell 2004).

Urbanization also exerts selective pressures on traits that underlie species interactions (e.g., foraging traits and defense traits), driving changes in community dynamics that control ecosystem functions. New selective forces in urban environments can alter the population dynamics of predators and reconfigure trophic interactions between predator individuals and their prey and the flux of organic matter in ecosystems (Terborgh and Estes 2010).

Mapping Phenotypic Evolution to Urban Ecosystem Function

Scholars have documented how urbanization affects primary productivity, nutrient cycling, hydrological function, and biodiversity through direct and subtle changes in climatic, hydrologic, geomorphic, and biogeochemical processes and biotic interactions (Pickett et al. 2011; Shochat et al. 2006). Simultaneously, scientists are increasingly interested in studying contemporary evolution in urban ecosystems (Donihue and Lambert 2014; Marzluff 2012; Partecke 2014). By explicitly linking urban development to heritable traits that affect ecosystem functions, we can start to map the eco-evolutionary implications of human-induced trait changes for those species that play an important functional role in communities and ecosystems, and we can identify gaps in existing knowledge (table 6.2) (Loreau and de Mazancourt 2013). The scale at which species perform different ecosystem functions could be a key to understanding relationships among urbanization, functional diversity, and ecosystem stability (G. D. Peterson, Allen, and Holling 1998).

Co-Evolutionary Dynamics in Hybrid Ecosystems

Cities evolve through a complex series of interactions involving a vast number of components, agents, and decisions. I hypothesize that co-evolutionary processes can have a significant impact on the evolutionary process and the adaptive capacity of coupled human-natural systems because of the potential for change and innovation. Humans provide opportunities for evolutionary change (Donihue and Lambert 2014; Hendry and Kinnison 1999), but human activity can also facilitate reverse speciation by inhibiting the process of species divergence (Seehausen 2006). Eco-evolutionary feedbacks are mediated by co-evolutionary interactions between species or genes; these can include strictly genetic co-evolution or gene-culture co-evolution (Durham 1991).

Expanding the spectrum of phenomena that can cause evolutionary change to include developmental bias and niche construction would more effectively represent the complex dynamic of interactions that take place between co-evolution and eco-evolutionary feedback in a human-dominated world (Laland 2014). A significant role is played by culture variations in phenotypes acquired directly and indirectly through social learning. Many species of mammals, birds, fishes, and insects have learned novel behaviors, such as diet modification, foraging skills, and antipredator behavior (Laland et al. 2011).

In cities, completely novel interactions between human and ecological processes might produce novel ecological conditions and unprecedented expressions, leading to new ecological patterns, processes, and functions. An example is the interaction between seed dispersal and road transportation in urban environments. As humans become agents for dispersing seeds, they facilitate competition among species, helping to determine which species thrive. Meanwhile, the structures and infrastructure that humans build provide new vectors and pathways for seed dispersal and new habitats that determine species' survival. Vehicles alter the mechanisms and patterns of seed dispersal in urban areas, making it far easier for nondispersing seeds to spread, sometimes farther than five miles (Von der Lippe et al. 2013). Humans also provide unintentional novel habitats, such as abandoned rail corridors and vacant lots.

Eco-Evolution and Innovation in Hybrid Ecosystems

Urban ecosystems are not simply complex coupled human-natural systems (Liu et al. 2007a). They are hybrids (Alberti 2008)—and that fact has an enormous impact on the eco-evolutionary dynamics of coupled human-natural systems. It is their hybrid nature that makes them unstable and unpredictable but also capable of innovating (Bettencourt 2013), allowing coupled human-natural systems to co-evolve and change (Allen and Holling 2010). Novel interactions in urban ecosystems might

trigger unprecedented dynamics and unpredictable changes, with significant implications for system functions and dynamics (Scheffer et al.
2012). At the same time, novelty in hybrid systems is a key component of
reorganization and renewal. Complex systems provide multiple possible
solutions for a given evolutionary problem (Wagner 2005). In a hybrid
system, that capacity enables the system to accumulate differences and
still maintain preexisting functions. In genetics, this capacity is a fundamental source of novelty. Novel phenotypes in interspecific hybrids
emerge from the interactions of two divergent genomes (Landry, Hartl,
and Ranz 2007). Despite the emerging divergence, their molecular co-
evolution ensures that their functions are maintained.

In these tightly coupled systems, opportunities for resilience and
adaptation emerge from the inherent uncertainty of complex cross-scale
human-environment interactions, which vary in both space and time. In
studying genetic networks in biological systems, Torres-Sosa, Huang, and
Aldana (2012) found that critical systems exhibit important properties
that allow robustness and flexibility: quick information processing,
collective responses to perturbations, and the ability to integrate a wide
range of external stimuli without saturation. Such interactions in urban
systems are highly influenced by technology and infrastructure (Anderies, Janssen, and Ostrom 2004). A key factor governing these interactions might be the lag time between human decisions and their impact,
which could be delayed and distributed over long distances (Liu et al.
2007a), because it regulates the relationships between humans and natural resources through both physical and social mechanisms.

Divergent scenarios may emerge from the complex dynamics of
hybrid ecosystems undergoing critical transitions. Figure 6.3 represents
three hypothetical scenarios of change: ecological, evolutionary, and eco-
evolutionary. In chapter 2, I described how, as a region urbanizes, ecological processes supporting the urban ecosystem may decline (figure 6.3a)
until they reach a threshold and drive the system into an unstable state
(cf. figure 2.14). Under an *ecological change scenario* (figure 6.3b), the system shifts into a new state in which ecological processes are highly compromised and drive the system to collapse (Alberti and Marzluff 2004).
Evolutionary change implies changes in the distribution of phenotypical
traits that may lead to a faster and earlier decline in system function or
delay and/or buffer system shifts, depending on the affected functional

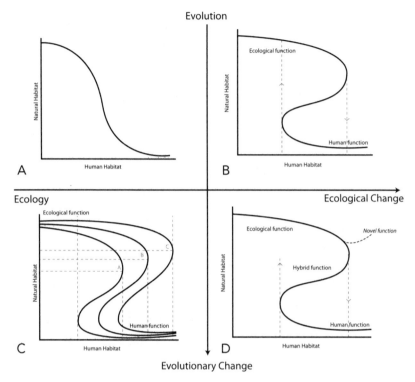

Evolution

Ecology

A — Human Habitat / Natural Habitat

B — Human Habitat / Natural Habitat — Ecological function / Human function

Ecological Change

C — Human Habitat / Natural Habitat — Ecological function / Human function

D — Human Habitat / Natural Habitat — Ecological function / Hybrid function / Novel function / Human function

Evolutionary Change

FIGURE 6.3a–d   Hypothetical scenarios of eco-evolutionary change. (*a*) As a region urbanizes, ecological processes supporting the urban ecosystem decline. (*b*) Under an ecological change scenario, they may reach a threshold and drive the system into an unstable state and eventually drive it to collapse. (*c*) Evolutionary change implies changes in the distribution of phenotypical traits, which may lead to a faster and earlier decline in system function or delay and/or buffer system shifts, depending on the affected functional traits. (*d*) Under an eco-evolutionary scenario, a novel system dynamic may emerge.

traits (figure 6.3c). Also, we do not know whether the adaptation of eco-logical processes to urbanization could make the urban ecological system more resilient to urban stress. I hypothesize that under an *eco-evolutionary scenario*, a novel system dynamic may emerge (figure 6.3d.)

Although evolutionary biologists have recognized that interactions in hybrids are a significant source of innovation in co-evolutionary processes, most researchers have seen the hybrid nature of urban eco-systems as a threat to ecosystem stability and resilience. In contrast, my hypothesis H4, above, suggests that hybrid ecosystems can be a source of innovation in eco-evolutionary processes, as hybrid mechanisms are essential to maintaining ecosystem functions while simultaneously allowing systems to co-evolve and change. Understanding the bases of these newly generated interactions is central to understanding co-evolution and adaptation in hybrid systems.

Concluding Remarks

From a planetary perspective, global urbanization could signal another critical transition for the evolution of Earth (Hughes et al. 2013). For most of its history, Planet Earth has experienced long periods of relative stability dominated primarily by negative feedbacks. But the recent increase in positive feedback (e.g., climate change), and the emergence of evolutionary innovations (e.g., novel metabolisms) (Lenton and Williams 2013) could trigger transformations on the scale of the Great Oxidation (Goldblatt, Lenton, and Watson 2006).

The increasing complexity and interdependence of socioeconomic networks and rapid tele-coupling can produce "tipping cascades" in the Earth's system that lead to unexpected regime shifts (Gaertner et al. 2014; Helbing 2013; Hughes et al. 2013). Only a formidable collaboration among scientists can address major questions such as these: What role do humans play in the evolution of Earth? Can the emergence and rapid development of cities change the course of Earth's evolution? Might different patterns of urbanization alter the effect of human action on eco-evolution? Can urbanization determine the probability of crossing thresholds that will trigger abrupt change on a planetary scale? How can researchers tackle such questions in new and productive ways?

I have argued that to address these questions, we must consider several levels of human interaction with ecological and evolutionary processes. First, what influence do humans have on population dynamics and community assembly? Second, how do human settlements influence heritable trait changes that support ecosystem functions? Third, we should expand the notion of eco-evolution to consider both the genetic and cultural co-evolution of human-natural systems (Laland 2014; Marzluff 2012). And a fourth level of inquiry should focus on how human-driven eco-evolutionary feedbacks affect ecosystem stability and regime shifts.

By integrating humans into the study of eco-evolutionary feedbacks, ecological scholars might be able to reconcile key theoretical concepts, including niche construction and community assembly, and redefine Hutchinson's (1957) "realized niche" on an urbanizing planet (Alberti et al. 2003). Doing so could also resolve important puzzles in island biogeography and explain contradictory empirical results (Marzluff 2005; Shochat et al. 2010). Rethinking evolutionary processes in a human-dominated world also implies expanding the notion of evolutionary causes beyond those that directly change gene frequencies to acknowledge reciprocal causation between selection and environmental changes and the significant role of phenotypic plasticity (Endler 1986; Laland 2014), including characteristics that organisms acquire through social learning (Laland et al. 2011) in directing evolutionary change (Laland 2014). I contend that studying how humans mediate eco-evolutionary feedback through urbanization can contribute significantly to progress and synthesis in evolution and ecology.

7

Reverse Experiments

TO DEVELOP AND TEST A THEORY OF URBAN ECOLOGY AND THE ROLE that cities play on a planetary scale, we need to redefine research methods and experiments and rewrite the protocols for collecting and synthesizing data. Several methodological challenges have become evident in the study of urban ecosystems: the complex dynamics and multiple confounders in determining causal effects, the difficulty of generalizing across regions and scales, the mismatches of scale across human and ecological system domains, the lack of predictability and certainty, and the problems of defining reference conditions for coupled human-natural systems and of quantifying human well-being. The unprecedented availability of detailed data, increased computing capability, high-resolution real-time sensors, and widespread mobile communication offers unique opportunities to meet these challenges. This chapter discusses the idea of designing studies as reverse experiments through which we can learn how urban ecosystems function, evolve, and succeed.

Inverse Problems

Studying urban ecosystems poses new challenges to ecology. The complex interplay between natural and human systems is one such challenge. A major problem is that most of the variables and interactions driving ecosystem function in urbanizing regions are not known. Even more challenging is the fact that scientists have yet to define and describe what constitutes a functioning urban ecosystem.

Urban ecosystems are complex coupled human-natural systems in which organisms and communities interact according to mechanisms and rules not fully explained in ecosystem ecology. Evidence from an increasing number of studies indicates that the assumptions of traditional theories in ecology (e.g., the disturbance hypothesis, in which humans are not included as agents) and social science (e.g., the idea that rational agents interact through efficient and stable markets) do not hold. To understand such systems, it may be necessary to both revise concepts of ecosystem function, stability, and optimality and develop new definitions of the ecosystem concept (O'Neill 2001), its dynamics (Pickett et al. 2011), and evolutionary feedbacks (Alberti 2015). We must redefine our research approaches, methods, and experiments.

In urban ecology, we face what geophysicists call inverse problems. An inverse problem is one whose initial inputs are a collection of observed measurements that are used to infer a model (or models) of the governing system that generated the outcomes; this model is the solution to the inverse problem. To explain the nature of inverse problems, the Russian mathematician Sergey Kabanikhin (2008) pointed to the way our brain uses previous experience to reconstruct an image by interpolating the limited information our eyes provide. Similarly, if we have accumulated prior information through experience, we are far less likely to make errors in interpreting and resolving a problem. However, when we attempt to understand complex phenomena or solve problems we have not previously encountered, the probability of error increases; hence, the solution is unstable. These are what mathematicians call ill-posed problems: problems for which more than one solution may exist (Sabatier 2000).

Inverse problems are intrinsic to urban ecology. Over the centuries, human societies have experienced and learned to deal with new challenges (e.g., access to clean water, poor sanitation, and air pollution) posed by the transition to urban life. But the scale and pace of current urbanization are unparalleled in the history of humanity and Planet Earth. Most current problems that local communities face in cities are new for humanity. Equally unprecedented are the scientific and policy challenges that the emergent problems pose, especially in the face of rapid climate change. What makes urban communities resilient to extreme climate events? How can we best prepare for, and mitigate, potential impacts? How can we adapt to new conditions?

Inverse methods aim to reconstruct phenomena that are difficult to observe or measure directly but which can be inferred from available observations. Scientists have documented several examples of regime shifts in ecosystems, such as the transition from a coral reef to an algae reef, from a tropical forest to a grassland, or from clear water to eutrophication. We still do not fully understand the emergence of regime shifts and their potential effects on urban ecosystems. What, for example, are the processes through which biogeochemical and human activities, coupled with the built infrastructure, lead to urban water eutrophication or catastrophic flooding events?

In studying the Earth, geophysicists seek to determine a continuous function of the space variables representing the Earth's properties with infinitely many degrees of freedom, as Snieder and Trampert (1999) pointed out. Yet, real experiments can result in only a finite number of measurements, which implies a finite data space: mathematically, the solution is not unique. In urban ecosystems, complex patterns and processes, multiple levels of organizations, and multiple causes of phenomena are intrinsic characteristics. History plays a significant role: such systems may be simultaneously contingent on their past state and on current conditions (Gould 1980). Thus, experiments designed to tease out a single cause may not be appropriate (W. C. Thompson 1989).

Studying an inverse problem is particularly complex given that the solution is unstable, and small errors in data cause large changes in inferred explanation (G. King 1997). Furthermore, investigators make assumptions about sampling, measurement error, and probability distributions to approximate a solution that may lead to divergent results (Biondi 2014). As compared to well-defined problems, which may be solved by using a set of preestablished operations, the processes involved in solving inverse problems are far more complicated and challenging (Hayes 1989).

In complex systems, causality is an inverse problem: that is, multiple causes may produce the same effect, so we cannot rely on traditional inferential approaches. To solve an inverse problem may require using observations to infer the values of some parameters. But observations should be used only to determine that possible solutions are false, not to deduce any particular solution (Tarantola 2006). In solving inverse problems with multiple possible solutions, observations can help us to identify a set of possible causes or forward model solutions, rather than a best solution, and to develop a scheme of multiple testable hypotheses

(Gomez-Ramirez 2013). Using a Bayesian approach, which relies on prior knowledge to predict future events, Tarantola (2006) suggested that we can start from prior information to sequentially create an infinite number of models to be formally tested using new information.

Inverse thinking aims to uncover the underlying principles governing a complex system and to build robust models in the face of uncertainty and complexity (Gomez-Ramirez and Sanz 2013; Tarantola 2006). A Bayesian approach allows us to deal with uncertainty in inverse problems by using new information to update the probability distribution for the variable of interest from prior to posterior (Gomez-Ramirez and Sanz 2013; Iglesias and Stuart 2014). Monte Carlo–Markov Chain (MCMC) methods are a powerful tool for applying this approach to ecosystem modeling. However, we may need to develop specific MCMC algorithms in order to use Bayesian inverse methods with complex ecological models that have large numbers of parameters (Dowd and Meyer 2003).

The idea of starting from data to build models does not imply that urban ecology should rely exclusively on inductive reasoning. It simply establishes a starting point for developing an iterative process that more effectively combines inductive and deductive reasoning. In a recent article in *BioScience*, Marquet et al. (2014) suggested an essential first step toward achieving a new level of integration in ecology: we must develop greater balance between inductive and deductive approaches. Ecological research has been dominated primarily by inductive approaches, a tendency exacerbated in recent years by the rapid development of technologies and big data (e.g., the availability of complete genomes and metagenomes). Expanding the role of theory and developing "efficient theories" (Marquet et al. 2014) are both critical to advancing integration and achieving synthesis in ecology.

Designing Experiments

Much debate is emerging in urban ecology about the nature of experiments and the challenges to research design. The hybrid nature of urban ecosystems makes them inherently unstable and uncertain, which leads to fundamental questions: What governs stability and resilience in urban ecosystems? How can we design an experiment to address this question?

A possible strategy for rethinking the role of experimental studies in urban ecosystems is to build on the concept of inverse problems and inverse modeling to design reverse experiments. While most experiments start with a research question and testable hypotheses about how the world works, a reverse experiment starts with the real world and asks, "What do I see?" or "What works?"

In principle, real-world observations provide us with clues about the structure and function of a given system. Scientists use these clues to conceptualize observed phenomena and organize available data to both explain existing observations and make future predictions (Marquet et al. 2014). But conceptualizations do not occur in a vacuum. In studying urban ecosystems, we build on concepts and theories of ecology. And yet assumptions regarding key properties (e.g., spatial and temporal heterogeneity) of ecological systems will need to be revised to account for the dynamic interactions between human and natural systems as they occur in cities. Furthermore, the dynamics of urbanizing regions depend on multiple processes operating over multiple scales. This fact has significant implications for research design.

The stability and resilience of urban ecosystems are difficult to study via an experimental approach. Studying regime shifts and system resilience implies relying on reconstructions and proxy records; in such studies, incomplete information is used to infer the underlying parameters and processes governing system dynamics. In fact, only a few aspects of stability or resilience have been studied experimentally or quasi-experimentally, and theories and experiments are often mismatched, so the tests are inconclusive (Ives and Carpenter 2007).

Inferring the future of urban ecosystems poses significant challenges to scientific research (Coreau et al. 2010). Science requires that hypotheses be formulated and tested through experiments (Aligica 2003). But experiments cannot be conducted in future studies since patterns and processes cannot be observed. Even when we rely on past observations, we must draw an important distinction between observational and experimental research (Biondi 2014). In observational studies, investigators cannot randomly assign treatments to subjects; therefore, the potential differences in starting conditions lead to potential bias in the estimates. This is known as Simpson's paradox (Pearl 2009).

Several factors challenge our ability to predict future conditions in coupled human-natural systems (Coreau et al. 2009). In a recent review of the literature, Coreau et al. (2010) identified some key challenges of predictive modeling and emerging approaches. The first is the complexity of coupled human-natural systems: their nonlinear dynamics, including threshold effects, spatial and temporal heterogeneity, and/or interactions between scales and processes (Carpenter and Brock 2004; Liu et al. 2007a). Also, studying complex systems often implies sample sizes too small for statistical generalization of probabilistic behaviors. And it may not always be realistic to rely on historical data and assume that current dynamics and relationships hold across time. Finally, uncertainties about drivers of changes can have major consequences for the possible futures of such systems (Carpenter 2002).

The key role of spatial heterogeneity in maintaining ecosystem function has long been recognized by landscape ecologists who challenged the traditional assumptions of homogeneity made in previous ecological studies (M. G. Turner 1987). In urban ecosystems, key properties, such as heterogeneity and connectivity, pose new challenges to those designing experiments, as both these properties have multiple sources and vary with the scale of observation. Yet such complexity cannot be disregarded since it has critical implications for the functioning and stability of such systems (Roff 1974).

Consider, for example, the heterogeneity of patch types, shapes, and sizes resulting from intermixed land uses in urban regions or the reconfiguration of resource flows and natural networks emerging from the development of built infrastructure. Despite the increasing emphasis on spatially explicit approaches to characterizing natural environments, most ecological experiments in human-dominated systems assume that space is homogeneous when they characterize the human component within the spatial social unit in order to focus primarily on emergent properties. And yet it is that internal heterogeneity, or the interplay among different heterogeneous elements, that controls system function and change.

The sources of heterogeneity and its emerging patterns in urban ecosystems also vary with scale (Pickett et al. 2008). At a finer resolution, land cover, patch type, form, and size are controlled by household preference and may reflect the demographic heterogeneity of households across diverse neighborhoods. At the city or regional scale, they may be controlled by collective behavior that drives choices about transportation or storm-water infrastructure. Furthermore, we only partially understand how human sources of heterogeneity interact with ecosystem functions in human-dominated systems, which poses additional challenges to experiment design and suggests that an exploratory approach and advanced sensitivity analyses may be more appropriate.

When we consider how human processes alter key properties of ecosystems, we see the need for important changes in both ecosystem theory and methodology.

Success Stories

Building on the concept of inverse problems and inverse modeling, we could use "success stories" to start to define *function* and *stability* in urban ecosystems. An example is seen in the study of resilience and system shifts. The design of a reverse experiment starts with the real world and asks, "What characterizes robust systems?" Resilience studies rely on the assumption that we know how ecosystems and societies maintain their functions in a given state and prevent themselves from moving into an alternative state. When a system shift occurs, or when what we expect to work does not, we define and test hypotheses about what failed, beginning with the assumptions we have made about how the system works. We design experiments with the assumption that we know how a system works and that we know its drivers, controlling variables, and boundary and reference conditions.

But when we are dealing with complex and relatively unknown hybrid systems, it is most likely that we have not identified some (or many) of the important elements at work. What if we are missing some fundamental rules or the most general principles governing the targets of our investigation? To understand resilience in urban ecosystems, a reverse experimental approach may prove more appropriate to tackling questions and could provide a road map. Such an approach could enable investigators to address questions about a broad range of topics—from the resilience of a specific system infrastructure supporting essential urban services (e.g., water supply, wastewater, and power) to the resilience of local communities.

As examples, consider the following: What enabled the community of Cedar Rapids, Iowa, to respond effectively to a vast flood, or residents of

New York and New Jersey to respond to the devastating Hurricane Sandy? What explains the capacity of communities such as New Orleans to recover from dramatic and devastating natural disasters? Are there general system properties that explain system and community resilience?

To begin to address such questions, we can articulate testable hypotheses regarding fundamental properties and principles governing the dynamics, stability, and evolvability of complex systems and networks (e.g., biological, ecological, and economic systems). As I discuss in chapters 4 and 10, studies of complex systems are revealing that systems that are more heterogeneous and modular tend to be better able to adapt compared to those whose elements are highly connected and homogeneous (Scheffer et al. 2012). Other properties of complex systems that have been posited to enhance adaptive capacity and innovation are cross-scale interaction, early warning mechanisms, and self-organization (Walker et al. 2004).

Do these properties explain community response and resilience to extreme events? Traditional design, whether experimental or quasi-experimental, requires that we observe randomly selected communities that vary with respect to these four properties over time, and/or that we have records of their response capacity before and after disaster events. Yet this approach implies that we can accurately describe key system structures and dynamics. An alternative approach would ask the same questions by exploring patterns, variability, and discontinuities that can be detected at multiple scales of space and time across communities that vary in their responses to disasters and their resilience to extreme events.

Institutional structure can play a significant role in determining the ability of communities to adapt to change. The ability of institutional capacity to change and adapt over time may be even more significant in socioecological systems. Empirical evidence reveals that a community's capacity to adapt to a new system state through increased resilience or transformation is not a short-term process; rather, it emerges over the course of years. Historical and present examples of societal planning and action in response to urban environmental stresses provide opportunities to analyze connections between resilience and transformation. Major success stories result from proactive organizations, both formal and informal, that have been able to anticipate critical transitions and expand the reference time frame of decisions.

Communities' ability to effectively incorporate uncertainty into decision-making is a key component of institutional adaptive capacity. It is possible only to predict some potential regime shifts that can result from the interaction of uncertain climate change trajectories and urbanization. Emerging research on early warnings points to signals (such as critical slowing) within complex systems that can be read to detect looming thresholds and regime shifts, but our current understanding and monitoring of potential warning signals are extremely limited. However, many communities have succeeded in adopting policies, routing investment decisions, and implementing strategies to anticipate natural disasters and reduce their vulnerabilities.

London, for example, has pioneered planning for adaptation to climate change through several innovative strategies. Since 2002, the London Climate Change Partnership has brought together experts in the environment, finance, health, development, housing, government, utilities, and communication. In 2002, the UK Environment Agency initiated the Thames Estuary 2100 Project (TE2100), intended to identify the next generation of strategic options for London and the Thames Estuary to manage escalating risks of tidal floods throughout this century. The TE2100 developed "decision adaptation pathways" (see chapter 10, box 10.1) to provide a flexible approach to accounting for the uncertainty inherent in climate change predictions. In October 2011, the city adopted its Climate Change Adaptation Strategy, a strategic framework designed to simultaneously enhance the quality of life in London and protect the environment.

Rotterdam is also among the first cities to develop and adopt a strategy to "future-proof" itself. As early as 2008, Rotterdam began exploring and designing a series of interventions to ensure that its water would remain safe and accessible and to keep the city infrastructure robust as the basis for urban development. At the same time, Copenhagen adopted drastic measures to cut CO_2 emissions and began positioning itself to become the world's first carbon-neutral city. In 2011, it also adopted measures to deal with extreme rainfall events.

The city of New York has led recent initiatives to build resilient strategies to counter extreme climate events. In response to Hurricane Sandy, New York launched a Special Initiative for Rebuilding and Resiliency (SIRR). The initiative is a major update of PlaNYC, Mayor Michael Bloomberg's major sustainability effort, which began in 2007 and is intended to address the critical interplay among an increasing population, climate change, and an aging infrastructure (see chapter 10, box 10.1). In June 2013, the city released a comprehensive report (New York City 2013) detailing a series of strategies that address coastal protection, buildings, health services, and critical infrastructure, at an estimated total cost of $20 billion.

On November 1, 2013, the city of Da Nang, Vietnam, adopted a new policy in its post-typhoon recovery support programs for damaged households. It requires that all new housing construction in the city incorporate key resilience principles, such as storm-resistant construction techniques. This success stems directly from the Storm and Flood-Resistant Credit and Housing Scheme in Da Nang City, a microcredit and technical assistance program aimed at developing storm-resistant shelters in vulnerable districts of the city.

Researchers claim that institutional change is required if communities are to manage the effects of climate change in general, and of sea level rise in particular (Biesbroek et al. 2013; Moser and Ekstrom 2010). But how can institutional innovation and change in coupled human-natural systems occur and relate to global change? This is not well understood. Numerous multistakeholder and multiscale efforts are developing strategies to help communities adapt to anticipated climate change. Agencies are setting targets, developing policies, readjusting interagency agreements, reviewing planning rules and processes, and funding projects whose successes hinge on interactions between uncertain future climate changes and equally uncertain institutional feedbacks.

Key questions emerge: What are the drivers of institutional innovation and change? How do scale-bound organizations address multiscale issues? Under which conditions can stakeholder processes best support the development of alternative institutional designs that facilitate effective adaptation, including more cooperative ways to manage commonly pooled resources? Building upon evidence from successful examples of institutional responses, we can begin to articulate and test hypotheses about the mechanisms that strengthen institutions' capacities to adapt to climate changes through careful modifications of institutional, legislative, and regulatory controls.

Big Data and Emerging Technologies

In principle, the changing nature of available data, monitoring instruments, and computing capacity should yield increasingly sophisticated designs and experiments. In particular, we can more realistically represent complex processes and interactions among a variety of variables by reducing the historic problems of oversimplification, inappropriate assumptions, and imprecise calibrations based on noisy data. Although improvements in data can only partly address central issues in experimental design, big data and emerging technologies do provide new opportunities to use data-mining techniques and conduct retrospective analyses that could better support inverse modeling. But data answer questions that we are able to formulate. They cannot answer questions that we are not asking.

A new paradigm in ecological science is necessary to address some of the challenges that urban ecosystems pose. Pickett, Kolasa, and Jones (2010) saw the emergent philosophy of science and its pluralistic view as an opportunity to advance ecological understanding by clarifying the goals of science, the role of theory, and the practice of scientific research. Insights from contemporary philosophy offer a new perspective for integrating multiple viewpoints and address key gaps in ecological understanding that result from the development of separate subdisciplines. Furthermore, the combination of advances in data and the development of new tools provides novel opportunities for such integration.

Urban Sensors and Observatories

Sensory perception allows living organisms to engage their environment—to perceive threats and discover resources—and thus to survive. Reverse problems require sophisticated sensors. Recent advances in high-resolution, mobile, ubiquitous sensor equipment suggest promising means of accelerating our capacity to observe urban systems and their functions by collecting and processing large amounts of highly detailed data in real time. But utilizing such tools will require us to develop advanced analytical methods that incorporate elements of machine learning, data mining, Bayesian statistics, social networks, and other experimental approaches.

Novel urban scale monitoring systems need to be deployed to collect a variety of data at the appropriate space and time scale to realistically represent the complexity of urban structures and processes. Advances in remote-sensing technologies and development of high-resolution data provide unprecedented opportunities to answer complex cross-disciplinary questions about urban systems. High-resolution nighttime light images from remote sensors could dynamically and precisely monitor boundary changes of urban built-up areas. Globally available sources are the nighttime light (NTL) data collected by the Defense Meteorological Satellite Program/Operational Linescan System (DMSP/OLS) and the Visible Infrared Imaging Radiometer Suite (VIIRS) onboard the Suomi National Polar-Orbiting Partnership (Suomi NPP).

The ability to sense potential threats or impending changes in the environment has evolved throughout human history, and our ability to anticipate adverse events has gradually improved. A well-known historical example is the use of canaries in mines to signal dangerous levels of toxic gases. New sensing technologies have the potential to provide a variety of early warnings for local communities. But they can also expand our sensory capacity from the perception of merely local and immediate threats to the recognition of larger potential threats associated with global and long-term change. Global and distributed networks of urban sensors and observatories are critical but not sufficient for establishing a robust infrastructure of urban sentinels. Urban ecology will need to build a new intellectual capacity and analytical framework in order to develop and execute powerful reverse experiments.

From Experiments to Practice

Real-world problems may require action in the absence of complete information. Thinking backward from problems to data can help us to filter and prioritize our data requirements and to design experiments that target emerging policy-relevant questions, thus facilitating the translation of scientific knowledge into practice. By starting with observed problems as they design their experiments, scientists may better inform planning and design as they test the efficacy of design and management strategies and thus narrow gaps between science and practice (Grose 2014).

Broadening the objectives of science and acknowledging the plurality of approaches to defining scientific questions does not imply that all scientific questions should emerge from our current understanding of problems. In fact, scientific discovery may very well revolutionize the way that societies define or understand problems. However, real-world problems provide unique perspectives—a set of lenses for defining scientific questions—and thus a direct path between a science of cities and city-building. The multiplicity of solutions inherent in inverse problems captures current understanding of the complexity and unpredictable behaviors of urban ecosystems that lead to emergent properties (ibid.). Furthermore, the city is an excellent laboratory in which to observe general properties of urban systems and to test theories of urban dynamics.

Defining urban problems, however, is hardly a trivial undertaking. To formulate questions relevant to the practice of city-building, a large and diverse group of agents must be involved and must share in defining the problem. This is a complex task, requiring shared language among multiple stakeholders and the resolution of conflicts that are likely to emerge. Yet a shared definition of the problem is key to establishing two-way communication between science and practice.

Alternative views of experiments can also strengthen the relationship between scientific research and the practice of urban planning and management. One such approach is joint fact-finding (JFF), an emerging strategy for building shared understandings of complex socioecological systems that consists of a procedure used for collaborative decision-making in public policy (Hanna and Slocombe 2007; Susskind, McKearnen, and Thomas-Lamar 1999). JFF informs science-intensive public disputes by bringing a scientific team into the collaborative process and simultaneously builds public credibility for its scientific findings. Ozawa and Susskind (1985) first described JFF as a way to mediate science-intensive policy disputes, and it has since received attention in the fields of conflict resolution, adaptive management, policy studies, and decision sciences (Andrews 2002; Ehrman and Stinson 1999; Ozawa 1991; Susskind and Cruikshank 1987). Its commonly stated goal is to produce agreed-upon information, regardless of the ideological or personal interests of particular analysts or stakeholders (Susskind and Zion 2002).

Another emerging approach that has been used in urban design and planning is to connect scientific expertise with practical urban design through design experiments. Felson, Bradford, and Oldfield (2013) described a series of projects that embed ecological research into urban and landscape design in order to use scientific knowledge effectively. Bringing scientists into the context of design may allow researchers to engage directly in managing and shaping urban systems (ibid.).

Urban decision-makers increasingly recognize that scientific evidence is pivotal to ensuring robust solutions to the complex problems they face. Simultaneously, scientists have begun to realize that if science is to play an effective role in decision-making and practice, they must engage directly and interactively with decision-makers and the public. While policy and science inherently differ in both aims and procedures, all parties appreciate the value of producing knowledge that can effectively support complex decisions to address emerging environmental problems, and understand that such knowledge production requires collaboration among scientists, policymakers, and other actors (Cash et al. 2003; Hegger et al. 2012; Van den Hove 2007; Van Kerkhoff and Lebel 2006). Such collaboration ensures that all parties gain ownership of the process of developing, selecting, and implementing design solutions and of planning strategies (Hirsch Hadorn, Biber-Klemm, and Grossenbacher-Mansuy 2008; Wiesmann et al. 2008).

This approach to joint production of knowledge has received positive attention in recent efforts to mitigate and adapt to rapid climate change. Diverse social actors provide unique and essential contributions to the process of knowledge production in order to support adaptation strategies that succeed (Ostrom 1996, 2009). Among the most critical tasks facing the diverse array of agents of knowledge production are to redefine scientific questions and to challenge the methods of research.

Box 7.1. Inverse Modeling

An inverse modeling approach is designed to model a system whose dynamics are unknown. In such a system, the order of cause and effect is reversed: the observer starts from the effects in order to identify the set of rules that govern the evolution of the system. Using, as an example, the study of resilience in urban ecosystems, we can identify four key steps for developing inverse modeling. An urban ecosystem's resilience to climate change (e.g., extreme climatic events) can be conceptualized as an inverse problem and modeled as follows:

1. Alternative system responses. The first step is to identify critical exposures to potential regime shifts that may affect urban ecosystem functions and system stability by examining documented cases of regime responses to extreme events and regime shifts (e.g., changes in hydrological regimes and eutrophication of urban waters). Mapping the exposures and vulnerabilities of different ecosystem functions to critical exposures will allow researchers to characterize alternative system responses.

2. Slow and fast variables. The second step is to transform the identified regimes into critical environmental conditions and to identify the key slow and fast variables governing responses (e.g., extreme flooding events and droughts). Using a series of modeling applications, inverse linkages between the established observations and a series of hypothetical mechanisms governing the resilience of urban ecosystems can then be established.

3. Model testing. The third step is to test the hypothetical mechanisms, using existing data and expert knowledge to upgrade the model by combining both prior information on urban ecosystem dynamics and existing knowledge (earlier measurements) about uncertain parameters.

4. Alternative scenarios. Finally, using a series of alternative scenarios, researchers can expand the boundary conditions of forward models to assess how hypothesized mechanisms might generate alternative trajectories and to what extent those trajectories might diverge from present conditions. ●

8

Incomplete Knowledge, Uncertainty, and Surprise

SCIENTIFIC KNOWLEDGE IS, INHERENTLY AND INEVITABLY, INCOM-plete. In this chapter, I explore the role that incompleteness, uncertainty, and surprise play in the evolution of scientific thinking. Tentative hypotheses, initial findings, and problem frames offer a false sense of certainty about what we know. The greatest challenge of scientific discovery is to unlearn what we have learned and to do so systematically and continuously. Sometimes, in decision-making, we face a paradox: while attempting to employ the best evidence to make informed, authoritative decisions, we ignore the inherent incompleteness of knowledge. By focusing on the unpredictable dynamics and the uncertainty of coupled human-natural systems, I challenge the myths about stability, opti-mality, transferability, and adaptability that characterize urban planning. These myths are not necessarily false or true, but are instead simply incom-plete. They represent partial explanations of how human-natural systems work. To face current challenges in urbanizing regions, a co-evolving paradigm may be more appropriate: a view that focuses on unpredictable dynamics in urban ecosystems and on strategies as experiments that help us to learn how cities work and evolve. I conclude the chapter by redefining the questions that can lead us in changing the planning paradigm.

The Paradox of Knowledge

It takes but a simple step beyond the familiar (our home, neighborhood, or city) to experience the incompleteness of our individual knowledge. It takes a little more imagination, however, to understand incompleteness in science. What we do not know may be both disproportionately greater than what we do know and qualitatively different from what we can expect. We cannot define or quantify what we do not know.

Why can this realization be important both in thinking about advances in scientific thinking and in framing the role of science in solving society's problems? The reason is that what we know influences the way that we think about what to look for and where to look for it. Whether or not we state it explicitly, we all have an idea of what "complete" or "perfect" knowledge should look like. This idea is often an abstraction, though, and is rarely connected to how the world actually works. At best, it is an expression of the relationship between what we need to know and a set of stated questions that such knowledge would answer. For example, it might provide a framework for compiling evidence about how cities affect ecosystem function.

But what if the questions we ask, although good ones, are incomplete? I propose that the questions we ask are strongly shaped by what we already know, and that so, too, is what we define as "unknown." To tackle such limitations, I propose a distinction between *ignorance*, defined as what we consider unknown—as Firestein (2012) discussed in his book *Ignorance*—and what I term *incomplete knowledge*. In a recent essay, the physicist Igor Teper (2014: n.p.) asked, "If the constitution of nature itself were changing in time, how would you know?" He concluded that we cannot assume that the future will resemble the past, based on past observations, without falling into what the philosopher David Hume (1777 [1748]: 115) called circular logic. Such scientific challenges have enormous implications for planning and decision-making.

Strategic decisions about the future—such as whether to invest in urban infrastructure, manage urban growth, and conserve natural resources—are based on our assessment of the past and our expectations for the future. To a certain extent, observations about the past define the "reference" and "boundary" conditions of the multidimensional space that shapes our predictive tools. How we think about the future has significant implications for the choices we make and the decision-making processes we apply. Traditional approaches to planning and management typically rely on predictions of probable futures extrapolated from trends observed in the past. But such predictions might not be adequate.

Incompleteness

Incomplete knowledge, uncertainty, and surprise all affect our decisions, and they all influence and interact with one another in their effects. Drawing a distinction among them can help us to clarify their roles and to tackle them as we make decisions. What helps us make good decisions is not having perfect knowledge, but acknowledging that we do not have it. In his book *Obliquity*, the economist John Kay (2010) suggested that in a world that we understand only imperfectly, we achieve better decisions when we approach a problem obliquely or indirectly. He noted that the problems we face are rarely completely specified and that the environment in which we tackle them contains irresolvable uncertainty. Problem-solving in an uncertain environment is an iterative process of continuous learning.

Part of the problem is that we know more than we can define or translate into formal expressions because our knowledge comes from multiple sources and because we can see to only a limited degree how our knowledge emerges from those sources. There is also a notable gap between our ability to perceive and our ability to synthesize and translate what we see into a set of rules. Often we cannot translate what we see into a reproducible decision-making model—a problem that reflects our unjustified trust in any one limited view of rationality. Though we know more and more about the important role that association plays in human cognitive processes by complementing rational thinking, our scientific and political institutions underestimate the power that this complementarity has in scientific discovery and decision-making (Scheffer 2014). The arts may be more effective in allowing us to simultaneously access multiple sources of knowledge. Art does not require us to explain the rules by which the work of art connects the dots. Instead it embraces incompleteness.

The recognition that knowledge is incomplete has several implications. First, our knowledge about our role in the environment cannot keep up with the rapid pace of the unknown impacts of our actions. Van der Leeuw et al. (2011) pointed out that the knowledge upon which we base our actions is disproportionately smaller than the unknown dimensions that our actions will affect. I suggest that part of the explanation for this mismatch is that our tools for understanding how the world works are inevitably incomplete and do not fully represent what we do know. Another part is that we cannot simultaneously access many sources of knowledge and synthesize or make sense of the multiple perceptions of what is real. Incompleteness of knowledge, however, is not the same as ignorance. Ignorance may include participation in an unconscious or deliberate action, such as ignoring a particular fact because of unintended bias or intentionally choosing to avoid it because it is inconvenient, unpleasant, or overwhelming. *Incompleteness* refers to the inevitable limitations constraining any individual or point of view as it seeks to represent the workings of the world.

Uncertainty

The future is more complex than any prediction we can make. But why? In part, this is true because of our limited ability to construct models that are accurate and/or precise enough to reflect the assumptions upon which they are built and the variability of the data that they are fed. We are adept at improving a model's accuracy and precision. Yet regardless of optimizations we achieve in our models, the future will surprise us. Again, why?

The key to a long view—to seeing beyond the near future and through the details of the present—lies in refining our capacity to connect the dots, to ask new questions, to access new sources of information, and to expand the methods of inquiry to include the imaginary. Predictive models are not based simply on what we know, but also on what we think we know. They are not only incomplete in their representations of how the world works; they also reflect our assumptions about what is missing. Their incompleteness has major implications for predictive modeling and long-term decision-making. Uncertainty is a dimension we can measure to determine the probability of a future event, but probability distributions are derived from the observed variability of known events. However, the probability of future events is not simply uncertain in a statistical sense. It is unknown. Here I do not argue that the future is unknowable. I argue that it is not completely knowable using our current knowledge and modeling processes. By assuming that we know what we do *not* know—that we know the probability distribution of given phenomena—we build models that tell us what we can predict with a given level of confidence that we cannot estimate.

In science, all predictions involve some level of uncertainty. Although the uncertainty may be great in some cases, scientists most often assume that it is quantifiable. Complex systems are inherently unpredictable, however, and thus their uncertainty cannot be completely quantified. A factor that makes models even more uncertain in coupled human-natural systems is nonstationarity, or the evolution of the parameters that define dynamics (Schindler and Hilborn 2015). In coupled human-natural systems, future dynamics depend on multiple drivers, such as population growth and climate change, that have very different degrees of uncertainty. The probability distribution of future predictions for coupled human-natural systems depends on the probability distributions of uncertain trends in such drivers and on their interactions. But since future driver distributions may be unknown, the uncertainty in such predictions cannot be calculated (Carpenter 2002).

In coupled human-natural systems, uncertainty and unpredictability can result from unexpected interactions among the driving forces and from the reflexive interactions between human behaviors and humans' anticipated knowledge of the environmental changes that may be brought about by their actions. In some cases, probabilities may be fundamentally unknowable. We cannot predict unexpected interactions among multiple uncertain driving forces, such as possible trajectories caused by interactions between uncertain climate changes and technological innovation. We do not know the probabilities of the impacts and feedbacks resulting from such interactions: the actual impact could very well fall outside the predicted probability distribution of model projections. As Steven Carpenter (2002: 2080) pointed out, "Even the uncertainties are uncertain, because we do not know the set of plausible models for the dynamics of the probability distributions."

Surprise

Thresholds are transition points between alternate states or regimes (Liu et al. 2007b). A regime shifts between alternate stable states when a controlling variable in a system reaches a threshold, modifying its dynamics and feedbacks (B. Walker and Meyers 2004). Subtle environmental change can set the stage for large, sudden, surprising, and sometimes irreversible changes in ecosystems. Regime shifts depend not only on the external perturbation that affects a system, but also on the size of the basin of attraction—a region in state space in which the system tends to remain (Holling 1973; Scheffer et al. 2001) (see figures 2.11–2.13). In systems with multiple stable states, gradually changing conditions may reduce the size of the basin of attraction around a state: what Holling (1973) defined as a loss of ecological resilience. It is typically described using the heuristic of the fate of a ball in a landscape of hills and valleys. As I represented in figure 2.11, a small perturbation or external event may be enough to cause a shift to an alternative stable state. However, this loss of resilience makes the system more fragile, in the sense that it can easily be tipped into a contrasting state by stochastic events.

Recent studies have provided empirical evidence that alternative stability domains exist in a variety of ecosystems, such as lakes, coral reefs, oceans, forests, and arid lands (Scheffer et al. 2001). B. Walker and Meyers (2004) described a database documenting thresholds in ecological and socioecological systems that drive system shifts (Resilience Alliance and Santa Fe Institute 2004). Complex feedbacks between natural and ecosystem processes and thresholds can generate regime shifts, but in coupled human-natural systems, we may not be able to see the effects that environmental change has on human function and well-being until ecological changes reach a certain threshold.

Regime shifts in ecosystems are difficult to predict (Scheffer and Carpenter 2003). However, increasing evidence shows that ecosystem dynamics become more variable prior to some regime shifts (Berglund and Gentz 2002; Brock and Carpenter 2006; Carpenter and Brock 2006). For example, by studying the variability around predictions of a simple time-series model of lake eutrophication, Carpenter and Brock found that a rising standard deviation (SD) could signal impending shifts about a decade in advance. Brock, Carpenter, and Scheffer (2006) explained how this statistical signal can occur in one-dimensional systems, and Carpenter and Brock (2006) showed that the variance component related to an impending regime shift can be separated from environmental noise using methods that require no knowledge of the mechanisms underlying that shift.

Myths and Paradoxes

As has been true for most of human history, breakthroughs in science often emerge from paradoxes and from the letting go of myths. Myths may very well be grounded in knowledge and observations, but they are only part of a much more complex story made up of knowledge and paradoxes. Paradoxes reveal assumptions and biases about how the world works, which emerge in the learning process. Tentative hypotheses and problem frames acquire false certainty and accuracy and become part of what we know, so the greatest challenge to scientific discovery is to systematically and continuously unlearn what we have learned. Unlearning is about abandoning the belief that what we have learned is accurate, complete, permanent, and transferrable.

Challenges

Interdependence between human and ecological processes in cities creates unprecedented challenges for planners and designers. At the same time, it provides them unique opportunities to build resilient cities—ones that can cope with environmental change in the long term. Some of our failures in managing complex human-ecological interactions in cities can be traced to our biases or to myths about nature. Holling (1973) told us that the ways that we view nature ultimately affect the ways that we study and understand ecosystems. Myths about how nature works lead to unverified assumptions about its processes and dynamics and then to inappropriate policies and strategies to protect and manage it. Urban designers and planners, for example, tend to assume that ecosystems are stable and that their processes and dynamics are relatively well understood. In the real world, however, ecosystems vary in time and space, and ecological change is subtle, rapid, and highly unpredictable (ibid.). To address the inherent uncertainty of coupled human-natural systems, we need to expose some common urban-planning myths.

Stability: Thresholds remain constant over time and thus are predictable.

Planners have long assumed that systems are stable: that they return to equilibrium when confronted with external disturbances. A *steady state* is a condition in which nature exists at or near a persistent equilibrium. The steady-state paradigm holds that disturbance can be controlled and that optimization is a strategy to achieve sustainable carrying capacities. Recently planners have begun to acknowledge multiple equilibria, yet they assume that thresholds between alternative states remain constant over time and system shifts are predictable.

But this isn't the case. Coupled human-natural systems may exhibit nonlinear responses to perturbations. There may be more than one stable regime. Both the position of a threshold along a driving variable and the depth of the basin of attraction can change. Resilience is a dynamic property. In coupled human-natural systems such as cities, reciprocal influences between system shifts may occur in both the ecological and the social systems (B. Walker and Meyers 2004).

A good example is the use of downscaled climate scenarios to make decisions. Climate scenarios are each based on a single emissions or climate model. Each scenario contains a set of assumptions about the trajectory of individual variables and associated uncertainties. Even when considering multiple climate scenarios, each based on different sets of assumptions, decision-makers cannot account for potential interactions among uncertain trajectories.

Optimality: There is an optimal resilient urban pattern and an optimal type of infrastructure.

The idea of optimality—that one can find the optimum among a set of alternatives given a set of conditions—is a direct consequence of the steady-state paradigm. Planners come to assume that for a given problem, an optimal solution exists. Decisions based on seeking the optimum assume that we can quantify risks. However, in the presence of irreducible uncertainties, we encounter multiple plausible futures whose relative probabilities are unknown. Decision-makers confronted with investment choices for dealing with floods or droughts, for example, are often misled by the confidence intervals associated with such predictions (Wilby et al. 2004). Although we can estimate the likelihoods (and confidence intervals) of environmental impacts given a set of assumptions, it is not likely that these assumptions will hold and that the future will turn out to resemble the "best-estimate" scenario. In fact, nonlinear rates of change and potential interactions among uncertain drivers may produce outcomes that more closely resemble outliers and low-probability events (Jones and Preston 2011; Reisinger, Wratt, and Allan 2011).

The farther into the future we look, the greater the uncertainty of our forecasts becomes—and even our levels of confidence about the magnitudes of uncertainties are likely to decrease as we gain greater understanding of the complexity of coupled human-natural systems. These facts have enormous consequences for decisions with long-term implications, since given our new understanding of risk, it might not be possible to adjust existing infrastructure or incrementally revise our responses. It may not be feasible or desirable to increase protection of existing land uses by building on infrastructure systems designed under outdated assumptions (e.g., adding height to a levee or raising the floor levels of dwellings) in response to each improvement in understanding. Robustness, rather than optimality, might be a more appropriate target for planning and decision-making under conditions of uncertainty.

Transferability: What is resilient in one region and at one scale is resilient in other regions and at other scales.

Many planning strategies are based on the assumption that what is resilient for a specific system function, at one scale and in one region, is resilient for other functions, at other scales and in other regions. In complex social-ecological systems, however, multiple regime shifts may occur in multiple biophysical (e.g., climate, hydrology, and biogeochemistry) and human (e.g., social, economic, and political) domains at multiple scales. Furthermore, in evolving systems, changes in scale influence resilience (B. Walker and Meyers 2004). Increasing the scale of urban systems may expand resilience by adding to the diversity of ecosystem types upon which they depend, but urban expansion may also increase the relative costs and impacts of maintaining urban activities on a larger scale. If planners focus on the resilience of a specific subsystem at a specific scale, that narrow view may cause the system to lose resilience on other fronts. To maintain resilience, we must focus instead on maintaining adaptive capacity and coping with uncertainty.

Adaptability: We can maintain resilience by adapting our current institutional frameworks.

Human and natural systems have evolved through change, adaptation, and extinction but are studied by researchers in separate domains. As a result, we do not fully understand how *coupled* human-natural systems evolve and what the limits of their adaptability might be. Furthermore, as novel functions and dynamics emerge in urban coupled human-natural systems, we may have to rethink both resilience and the way that we frame regime shifts. Emerging system functions may require transformation, which implies a regime shift toward a new desirable state, not simply adaptation.

Adaptation planners assume that we can maintain resilience by adapting our current institutional frameworks. Yet just as we cannot simply adjust or upgrade existing infrastructure to deal with flooding and other climate changes (by, e.g., strengthening flood-control structures), existing institutional settings may need to be reinvented by revising traditional assumptions of problem-solving from top-down control to self-organization. The theory of complex systems provides a solid foundation for testing hypotheses about system properties (e.g., self-organization) that enable coupled human-natural systems to change their internal structure to adapt and evolve in response to external circumstances. To create a co-evolving paradigm, we may have to reconfigure current planning frameworks and fundamentally transform our current institutions for managing cities.

Planning under Uncertainty

At the core of the challenges faced by cities across the world lies our inevitable uncertainty about global environmental change. Not only are extreme events uncertain; so, too, is the very measure of that uncertainty (e.g., the tails of the probability distributions).

Under uncertainty, each problem has more than one solution, and solutions may change with time as conditions evolve and as our perceptions and understanding change. New viewpoints may emerge as we see and formulate tentative solutions. Uncovering a new viewpoint often exposes new opportunities and unexplored resources for problem-solving.

Redefining Our Questions

If uncertainty and surprise are fundamental elements in decision-making, perhaps we have been asking the wrong question:

How can we minimize uncertainty about future changes in order to select the most optimal strategies for controlling them?

A more effective question may be:

How can we characterize uncertainty about future changes in order to inform development of the most robust strategies for planning?

The assumption of traditional planning is that we could achieve a perfect decision if we could

- eliminate uncertainty,
- remove differences,
- have complete knowledge,
- have plenty of resources, and
- achieve perfect coordination.

But probably under such conditions, there would be no need to make a complex decision, as only one logical course would exist. In the face of uncertainty, we need to revise these assumptions. We will have to

- embrace uncertainty to build robust decisions,
- build on differences to explore opportunities,
- use information to test what we know,
- manage resources effectively to maximize benefits, and
- transform redundancy into partnership.

This view focuses not only on unpredictable dynamics in ecosystems, but also on institutional and political flexibility. We need to reconfigure our problem-solving strategies. Instead of seeking to reduce uncertainty, decision-makers should aim to identify robust strategies that will be appropriate under a broad range of alternative futures.

Direct experience of the dramatic shift in extreme climate-event probability across cities worldwide has significantly influenced how decision-makers perceive and assess risk. New York City's response to Hurricane Sandy is only one of many examples of emerging efforts to fully integrate uncertainty into local decision-making by redefining "extreme" events as "normal" ones. Such a change in perspective calls for what Rosenzweig and Solecki (2014) described as achieving a meaningful nonstationary public policy and institutionalizing climate-change adaptation and resiliency.

Future policies and management practices in urbanizing regions will succeed or fail based on their ability to acknowledge and address the complexities and uncertainties of coupled human-natural systems. When policies aim to stabilize an ecological system or eliminate its variability, the inevitable outcome is collapse (Carpenter and Gunderson 2001). This is where scenario-building can be a valuable process: it allows decision-makers to explore possible futures and account for uncertainty and surprise.

Uncertainty and Planning across Time and Space

The relationship between uncertainty and decision-making is complex. There are many sources and varieties of uncertainty and many diverse types of decisions. Not all aspects of uncertainty are untreatable. In fact, several of the important variables driving change can be foreseen on a relatively short time scale. Even some aspects of complex phenomena such as climate change have a relatively limited degree of uncertainty compared to less predictable elements of the same phenomena and thus pose different challenges in decision-making. It is also important to discriminate among different kinds of uncertainty. Consider, for example, the difference between the uncertainty that arises due to limited scientific understanding and or/agreement on the many feedbacks governing how the Earth's systems operate and the uncertainty of predictions caused by imperfect climate modeling that is based on universally accepted scientific bases. We must also distinguish between scientific uncertainty and disagreement regarding the timing of appropriate mitigation efforts.

Cross-scale interactions among variables governing the dynamics of coupled human-natural systems add further dimensions of uncertainty.

Both human and natural processes operate and affect each other at multiple temporal and spatial scales. What is resilient at the local scale over a short time period might not be resilient at regional or global scales over the long term (Carpenter et al. 2001). Trade-offs among system functions, thresholds, and tipping points vary across scales, and tele-connections (rural-urban, local-global, and across cities) imply that change and adaptation also operate across scales.

Human civilization is characterized by an increasing capacity to subsidize resources across time and space. Technological development associated with urbanization and agricultural transitions has expanded human ability to maintain system stability (e.g., grow food or supply water) in one place and time scale at the expense of other places and longer time scales. Yet after a certain threshold in the supporting ecosystem is crossed, system resilience is compromised (Carpenter et al. 2001). Using the Dutch Deltaworks Program as an example, Chelleri et al. (2015) show that the evolution of flooding infrastructure reflects different phases of multiple approaches to achieving resilience. Overlap between adaptation and transformation perspectives suggests that multidimensional resilience strategies are essential when dealing with regime shifts.

Similarly, there are very different levels of complexity and scales of decisions. Stafford Smith et al. (2011) proposed that we classify decisions according to their lifetime, their incremental or transformational nature, and their scale. Decisions may have a short lead time and short-term consequences or a short lead time and long-term consequences (e.g., whether to build infrastructure or develop land). It is possible to reduce complexity and uncertainty by identifying interactions among a decision's lifetime, the type of uncertainty in the relevant drivers of change, and the nature of the options for responding to adaptation. Decisions with a short time scale can rely on predictive models, which can produce relatively accurate future scenarios. Decisions with longer lifetimes must deal with potentially wider arrays of possible futures—and that may require integrating a diversity of approaches.

Rethinking planning in a time of rapid planetary change requires that planners understand the implications of incomplete knowledge, uncertainty, and surprise in decision-making and formulate strategies that vary and evolve across time and space.

9

Scenarios

Imagining the Possible

THIS CHAPTER LINKS SCIENCE WITH IMAGINATION IN THINKING about the future. It explores how, by navigating through time, we can uncover our biases about what we know and explore new ways to integrate uncertainty into decision-making. I suggest that we can learn from the future, and to demonstrate this, I ask how the imaginary futures presented in chapter 1 challenge the assumptions of urban design and planning paradigms. Using a hypothetical example of the planning process that a city might employ to adapt to climate change, the chapter draws a road map and highlights key elements of a creative process for strategic foresight.

The Power of Surprise

Surprises happen in the present and change the way we think about the future. It is when an unexpected event shifts the focus of our attention and rearranges our priorities that we typically become most creative in solving problems. When we encounter unexpected circumstances, both as individuals and as communities, we tend to engage in creative actions, see a greater range of opportunities, and gain the ability to rapidly determine the most effective response. Although this capacity is widely experienced on an individual level in the setting of extraordinary events, if we look closely, we see that everyday life events in cities are also extraordinary and that creative capacity is a crucial element in the way that humans operating in uncertain conditions make decisions. The key here is the unexpected. Under such circumstances, we release both our thinking and our capacity to act from the long-held expectations and inevitable biases that bind them—and that constrain our imaginations.

Key qualities that surprise brings to life include a sudden change of viewpoint and the realization that there may be many viewpoints and temporal scales that redefine our priorities and options for action. Other qualities include curiosity, perception, and intuition. Surprise leads people to access underutilized sensors and untapped resources and to try a new focus, to narrow down the problem and figure out possible solutions. When faced with unexpected circumstances, we tend to be more open to envisioning opportunities than we might otherwise be; we are more adept at identifying multiple solutions to novel problems.

In a recent study, Stahl and Feigenson (2015) found that surprise plays a critical role in infant learning. Infants selectively explore and learn from objects that violate their prior beliefs by engaging in hypothesis testing that specifically reflects the observed violations. Here I suggest that novelty is the fundamental element of surprise that allows for change: it prompts instability but is also a key ingredient of innovation and renewal. Faced with novel conditions, humans and other organisms learn, evolve, and adapt.

Learning from the Future

It may seem counterintuitive to suggest that surprise is a key ingredient of effective planning. Yet the best way to test a plan is to see how it helps communities deal with surprise. So why not incorporate surprise into planning from the outset? Why not intentionally harness the intuitive power elicited by confronting the unexpected? How to best introduce uncertainty and surprise into strategic planning and decision-making is our next challenge. I started this book with the suggestion that through imagining the future, we can transform the way we live in the present. What do the four hypothetical futures visited by Max, the subway rider in chapter 1, teach us about planning?

In Glacial City, we see that change may occur as a dramatic shift in what a city has experienced since its birth, but we also see that climatic regime shifts do not necessarily imply the end of life. Some species may adapt successfully, and novel species may emerge. This view suggests that human agency may play a more subtle role in the future of Earth than either of the two archetypical and opposing possibilities of resilience and collapse, which may oversimplify and fall short of what is possible and desirable for the planet and the human species. At one extreme, they reflect both our fears and our overconfidence that we have the capacity to control nature; at the other, they reflect a naïvely optimistic vision of what technology and science might achieve. They do speak clearly about the limits of our collective imagination. Glacial City proposes that humans have the opportunity to lead transformation toward a desirable co-evolutionary outcome.

Empty City is an important dominant archetype in the history of planning, one that is still present in the institutional frameworks of most planning agencies and the mental models of most practitioners. It is a future that relies on the illusion of control. Despite the incontestable evidence that the strength and rigidity of control structures are failing and are, perhaps, even driving the system to failure, control often reemerges in new forms. For planners and designers, it may be difficult to fully appreciate the concepts of emergence and self-organization and to reimagine their roles accordingly. The Empty City reminds us that old paradigms challenge the process of renewal and reorganization until a new paradigm emerges.

Dream City probes the limits of our imagination. In every epoch, architects, planners, and engineers, as well as historians and philosophers, have dreamed about the city of tomorrow, from the Ville Radieuse (Radiant City) of Le Corbusier (1935) to Ebenezer Howard's Garden Cities (1902) and Frank Lloyd Wright's Broadacre City (1932). More recent dream cities may float (e.g., the Lilypad, by Vincent Callebaut) or be submerged (e.g., the "Ocean City" concept, called Syph, by Arup Biomimetics). Some are guided by a conscious intent to establish a new relationship with nature (e.g., the green cities and biophilic cities movements) or to use the most advanced technologies (e.g., smart cities).

Yet people are rarely the protagonists in such dreams. Instead they are users and spectators. In Dream City, the fantasies of architects, city planners, and technocrats have become real, but what about the dreams of ordinary people? Dream City leads the protagonist of the story to realize that our cities and urban communities are much more complex, dynamic, and abundant with possibilities than any mere idea of them could be—not because imagination has any real limits, but because once we can represent the imaginary city, it becomes part of reality. This complex reality and the possibilities that are inextricably linked to its uncertain future are what Hybrid City accepts. In Hybrid City, surprise is part of the normal distribution of events, and uncertainty and instability challenge all the city's living organisms to evolve and reinvent themselves in the face of change. They foster learning and innovation and provide opportunities for transformation.

The Challenge

In urbanizing regions, multiple steady and unstable states exist simultaneously (Alberti and Marzluff 2004). This has profound implications for decisions about how to best respond to environmental change. At first glance, human and natural functions in an urban ecosystem may seem to be operating independently, but in reality, they are highly coupled (Alberti 2008). Consider, for example, how the built infrastructure in an urban ecosystem modifies hydrological functions. As an area becomes urbanized, humans tend to replace natural hydrological functions with built infrastructure; doing so lets them control the water flow, extract and distribute water for human uses, and purify water before it returns to natural water bodies. In this process, urbanization decreases the number and quality of natural hydrological functions and replaces them with built infrastructure that supports human functions.

The infrastructure's ability to serve multiple uses depends on the size, availability, and recharge capacity of the clean water supply. But the decline of natural hydrological function may constrain that supply. As a result of human pressure, the coupled hydrological function (both human and natural) may decline as ecosystem functions, both local and global, are reduced. The functional form of the relationships among ecosystem function, human function, and urbanization depends on the specific ecosystem functions being considered; it may also depend on alternative future conditions caused by complex interactions among drivers of change (e.g., climate or technology).

There is another factor that makes it even more difficult to model interactions among drivers and to predict potential shifts in system behaviors: their effects are both cumulative and synergistic. In general, environmental disturbances have an important impact when the factors causing them are grouped so closely in space or time that they overwhelm the natural system's ability to remove or dissipate that impact (W. C. Clark 1986). In cities, human pressure on ecosystem functions may cross thresholds beyond which stresses may irrevocably damage important ecological functions. In most ecological systems, processes occur stepwise rather than progressing smoothly, and sharp shifts in behavior are natural (Holling 1973). These related properties of ecosystems require us to consider how functional interdependencies affect resilience in urban ecosystems under highly uncertain future scenarios.

Modeling the future of urban ecosystems is challenging not only because the systems are complex and the phenomenology and modeling approaches are diverse, but also because those who participate in developing strategic foresights conceptualize and treat uncertainty differently. Many factors can exacerbate uncertainty. For example, we may not sufficiently understand a given phenomenon, or we may make systematic and random errors or subjective judgments. Natural systems can change abruptly and in discontinuous ways, and characterizing the responses of the system function involves thresholds and multiple domains of stability (Carpenter 2002; Scheffer and Carpenter 2003). Because the knowledge of environmental systems is always incomplete and uncertain, surprise is inevitable (Brock, Carpenter, and Scheffer 2006; Holling 1996), which has significant implications for planning.

Strategic Planning and Foresight

Strategic planning has traditionally relied on projections and forecasts of future conditions to assess potential risks and alternative actions. But the increasing complexity and uncertainty of decision-making have challenged the effectiveness of such approaches. Coupled human-natural systems test our traditional assumptions and strategies for planning and managing natural resources and the environment (Liu et al. 2007a). The success or failure of many policies and management practices depends on their ability to account for the complexities and uncertainty of these systems. The emergence of strategic foresight and scenario planning to address rapid environmental change is a fundamental shift in thinking about the future: from a single probable future toward a range of plau-

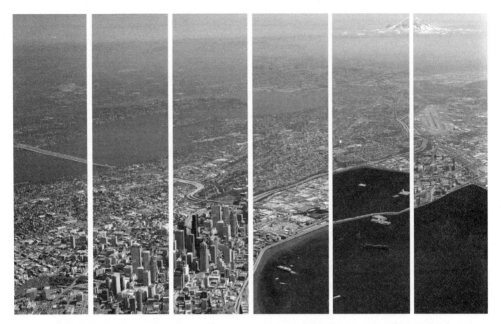

FIGURE 9.1   Representations of future scenarios for Seattle. Imagine Seattle fifty years from now. How will interactions among alternative development patterns, economic trends, and climate change affect ecosystem function and human well-being? Photograph: Aerolist-photo.com.

sible ones. From a theoretical standpoint, this is a shift in both the science and the practice of decision-making.

The objective of strategic foresight is better decisions, not better predictions (Dearlove 2002). Better decisions are *robust* under a range of plausible scenarios (figure 9.1). Such decisions may not be *optimal* under all plausible futures, but they will perform well across the range of plausible scenarios that might emerge (Sutherland and Woodroof 2009). Thus strategic foresight reconfigures the decision-making process (Cook et al. 2014): instead of first defining a set of objectives, it starts by framing the problem and characterizing the uncertainty in key driving variables. Focusing on the problem and the uncertainty, this process informs planners as they develop a range of plausible scenarios, prompting identification of new risks and opportunities. Perhaps the most important outcome of strategic foresight implementation is that it enables planners and managers to respond to surprise events with potentially significant impacts (Taleb 2007).

Strategic foresight identifies alternative scenarios as plausible sets of conditions from which to identify and select the most robust responses.

In contrast to traditional planning and design processes, a strategic foresight process systematically considers a range of plausible divergent futures, the hidden assumptions that underlie these futures, their potential consequences for policies and decisions, and the actions that might promote more desirable futures (Inayatullah 2005). Strategic foresight builds on what we know and simultaneously expands our mindset by enhancing our ability to keep the inevitable bias of the past from dominating how we think about the future.

Unfolding the Steps

How can we ensure that a process of creative foresight will be effective? The few practical experiments that have been documented in the literature suggest a few key elements, although researchers do not agree on a single "best" approach or offer evidence to support one approach or tool as superior to all others. The application of strategic foresight in environmental planning and urban design has not yet generated sufficient data to provide general principles. However, strategic foresight has existed since the 1950s and has been applied in a variety of fields and sectors, ranging from the military to business organization and governmental institutions (D. A. King and Thomas 2007).

Strategic foresight may involve a variety of approaches and tools (Cook et al. 2014) and is performed through several activities: (1) set the scope; (2) collect inputs; (3) analyze signals; (4) interpret information; (5) determine how to act; and (6) implement outcomes. Various approaches have been applied in different contexts, ranging from horizon scanning to scenario planning and backcasting (Godet 1986; Hines 2006; Robinson 2003; Sutherland et al. 2014), and some of them emphasize, and are generally more effective in performing, a particular step.

Setting the scope means first defining a shared problem—the focal issue—which establishes the limits of the system of interest. Then one should identify key issues and the important actors that must be involved in the foresight process. The focal issue of any foresight process guides each of the steps as well as the design of the exercise. In practice, the steps may involve a variety of approaches and tools (Cook et al. 2014).

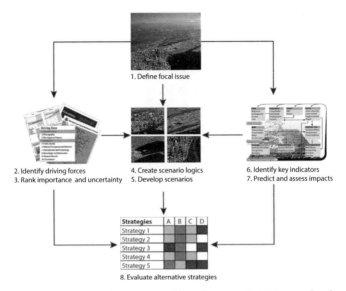

1. Define focal issue

2. Identify driving forces
3. Rank importance and uncertainty

4. Create scenario logics
5. Develop scenarios

6. Identify key indicators
7. Predict and assess impacts

Strategies	A	B	C	D
Strategy 1				
Strategy 2				
Strategy 3				
Strategy 4				
Strategy 5				

8. Evaluate alternative strategies

FIGURE 9.2   Scenario process and key elements. Participants develop a shared definition of the focal issue (e.g., ecosystem function) (1) based on both observations and expert knowledge to identify and rank key driving forces influencing future change (2) that are used to draw the scenario logics (3) and develop the scenario storylines (5). Indicators are selected to assess the impacts of scenarios (6) on the focal issue under predicted changes (7). Baseline scenarios and predictions are used to evaluate the efficacy and robustness of alternative strategies to address the focal issue (8).

Scenario planning is one of many approaches to strategic foresight. It involves eight steps: (1) set the focal issue; (2) identify key drivers; (3) rank the importance and uncertainties associated with the key drivers; (4) create scenario logics; (5) develop scenarios; (6) identify indicators; (7) assess implications; and (8) evaluate strategies (Alberti, Russo, and Tenneson 2013) (figure 9.2).

To demonstrate how scenarios can be used to assess alternative planning strategies, I provide an example of a hypothetical strategic foresight exercise designed to inform the development of a strategy for metropolitan adaptation to climate change (table 9.1). At the scale of a city or metropolitan area, such an exercise might involve a different scope, and thus a different set of variables and spatial and temporal scales, than would be appropriate for a regional plan for protecting biodiversity.

Table 9.1. Scenario planning elements, actions, and questions for a hypothetical city's plan for adapting to climate change

Scenario Element	Rationale and Objectives	Action and Sources	Adaptation Plan Example
Focal issue	Scenarios depend on the problem being addressed and the decision that needs to be made. *Develop a shared problem definition and conceptual framework to ensure that credible scenarios are developed.*	Based on a series of data, observations, historical documents, and expert knowledge, the scenario team identifies key drivers and processes and the relevant temporal and spatial scales at which they operate and interact.	• Which critical decisions face the city in the next 50 years? • Which key drivers of land use and climate change will affect the city ecosystem functions over the long term? • Who are the major agents of change and stakeholders? What are the objectives of the adaptation plan?
Driving forces	Driving forces are variables that influence the trajectory of the focal issues in significant ways. They might represent clusters of trends. For example, urbanization is a driving force that can affect both resource consumption and ecosystem change. *Identify key drivers of change affecting the focal issue.*	Based on the best available science, the team identifies the importance of different driving forces and characterizes their uncertainty.	• What are key drivers of change with respect to the focal issue? • What is the nature of the uncertainty (e.g., magnitude, variability, and/or direction of trend in a specific variable)?
Ranking importance and uncertainty	Interactions among the most important and uncertain driving forces determine the range of possible future outcomes. *Capture the most divergent yet plausible futures.*	Participants are asked to identify and rank driving forces based on their importance and uncertainty.	• Which key drivers of change are simultaneously the most important and uncertain with respect to the focal issue?

Scenario logics	Scenarios are hypotheses about plausible interactions that may take place given the uncertainty in key identified driving forces.	• Which major dimensions and endpoints best characterize the uncertainty? An example of a dimension for a climate variable such as precipitation is average rainfall, and endpoints may represent the magnitude.
	Define the organizing structure representing divergent yet plausible future conditions that can emerge from the interactions of uncertain driving forces.	
	The team uses selected driving forces to create frames for developing scenario logics. Frames are created by crossing the plausible trajectories of selected dimensions and endpoints of key uncertain driving forces.	
Scenario development	Scenario development entails researching and writing the narratives of each scenario based on the driving forces and identified agents.	• Which plausible futures emerge from interactions between a major or minor climate change outcome and alternative development patterns and/or infrastructures?
	Characterize four distinct and internally consistent stories of how the future can unfold.	
	Participants develop scenarios, story lines, and narratives. The process essentially follows the initial "what if" question emerging from a plausible hypothesis regarding the interactions of selected driving forces.	
Selected indicators	Indicators describe the implications of the scenarios for the focal issue (e.g., ecosystem function, resource availability, human health, etc.).	• What are the best (most relevant, available, or sensitive) indicators of ecosystem functions that capture the effect of divergent futures?
	Select indicators that are (1) relevant, (2) sensitive to scenarios, (3) quantifiable, and (4) communicable.	• Which warning signals can help us to anticipate critical transitions?
	The team uses scenarios to select key indicators that can effectively describe the current status and future impact of land use and climate change for ecosystem function and human well-being.	

(continued)

Table 9.1. (continued)

Scenario Element	Rationale and Objectives	Action and Sources	Adaptation Plan Example
Impact assessment	Scenarios identify potential risks that driver interactions may generate and expand the boundary conditions of existing model predictions. *Assess future conditions under alternative scenarios.*	The team uses predictive models to assess future trajectories of selected indicators. Scenarios are applied to expand the boundary conditions assumed in the models to make predictions.	• How can we quantify scenario outcomes as alternative future baseline conditions of ecosystem function? • How can models help us test the sensitivity of alternative strategies to future conditions?
Evaluation of alternatives	Scenarios aim to identify the most robust strategy given the plausible future. *Identify the most robust and effective planning strategy.*	Using potential risks and opportunities of alternative scenarios, participants explore possible strategies. Based on selected indicators, participants evaluate the efficacy and robustness of alternative strategies across the alternative plausible futures.	• How do we evaluate adaptation strategies? • What are potential trade-offs across strategies? • Which strategies best incorporate uncertainty? • Which warning signals can help us to anticipate critical transitions?

Once the scope has been defined, the next step in developing strategic foresight is to collect inputs: to solicit and compile information, from a wide range of sources, on past trends and potential future trajectories of selected drivers (Amanatidou et al. 2012). The foresight team engages in various iterations of data collection, expert interviews, and conceptualization. The diversity of knowledge and expertise among the participants ensures a robust outcome in the foresight process (Bengston, Kubik, and Bishop 2012).

The team then analyzes key signals by integrating data sources, examining drivers' trajectories and interdependencies, and modeling potential impacts. In the fourth step, interpreting information, the team examines the influence of assumptions and uncertainty in the alternative scenarios. This information constitutes the basis for determining a diverse set of possible actions, assessing alternative strategies under different scenarios, and developing a monitoring strategy.

Scenario Planning

Scenario planning offers a systematic and creative approach for bringing the future into present decisions by challenging our assumptions and expanding old mindsets (Alberti, Russo, and Tenneson 2013). Scenarios are narratives about alternative environments in which participants can play out their decisions about planning and management strategies (Schwartz and Ogilvy 1998). They are plausible accounts, based on empirical data, expert knowledge, and participants' experience of how relevant forces might interact.

Scenarios are not predictions or visions. Rather, they are hypothetical alternative futures that highlight the risks and opportunities involved in specific strategic issues (Schwartz 2005). They help us to see how robust alternative strategies may be plausible—or not—under future conditions (Schoemaker 1995; Wack 1985). As we begin to develop shared hypotheses of alternative futures, we can envision both the challenges and opportunities to come.

Scenarios are written as plausible stories—not probable ones. Traditional approaches to planning and management rely on predictions that are based on probabilities and quantified uncertainties. Peterson, Cumming, and Carpenter (2003) described a prediction as the best possible estimate of future conditions. As compared to scenarios, predictions assume knowledge of the probability distributions of specified ecological variables at specified times in the future. However, the accuracy of such a prediction depends on current conditions, specified assumptions about drivers, the measured probability distributions used in model parameters, and the measured probability of the model itself being correct (J. S. Clark et al. 2001). Alternative scenarios expand the boundary conditions of predictive models by acknowledging irreducible uncertainties.

Planning through New Lenses

Scenarios synthesize observations through new lenses developed by exposing the assumptions within our professed knowledge; by challenging assumptions, scenarios allow scientists to capture pictures of how the world could look. A set of such pictures helps decision-makers to make long-term plans by generating discussions around desirable and conceivable futures, and it can serve to expand the time scale of decision-making.

Empirical studies of the practices of scenario planning that reflect on the challenges and opportunities of this approach for decision-making are just beginning to emerge. In a recent study, Oteros-Rozas et al. (2015) drew insights and experiences from twenty-three case studies on the opportunities and limitations of using scenario planning for place-based social-ecological research. The authors concluded that scenario planning did succeed in engaging stakeholders, increasing dialogue, and resolving conflicts by building a shared understanding of the problem and encouraging a complexity perspective on social-ecological systems. In the following paragraphs, I draw a few insights from both the literature (van Vliet et al. 2012; Carpenter et al. 2015a; Oteros-Rozas et al. 2015) and my direct experience of scenario planning in the Puget Sound region (Alberti, Russo, and Tenneson 2013).

Focus on Resilience

How can scenarios help decision-makers shift their attention toward resilience? Resilience, as noted earlier in the book, is the capacity of a system to tolerate disturbance without shifting into a qualitatively different state that is controlled by a different set of processes. Resilience theory relies on assumptions about four interdependent elements of coupled social-ecological systems: complexity, change, diversity, and uncertainty. At times, maintaining or enhancing the resilience of one subsystem is costly, as it means reducing the resilience of another. Planning decisions may involve important trade-offs that cannot be eliminated but can be explicitly addressed through negotiations among various stakeholders.

Expand the Decision Context

Using scenarios to plan for the future implies that we can define present problems and identify drivers of change, actors, and their behaviors and views. The diverse and heterogeneous composition of cities, in terms of values, backgrounds, and experiences, leads to both innovation and conflict. The scenario-building process expands the decision context and provides opportunities to shift power domains (actors), conceptualize problems (information), and pay attention to the politics (priorities) and innovations (substitutable actions) that divergent strategies may imply. An expanded decision context helps us to explore strategies that are generally more equitable, flexible, proactive, and anticipatory.

Challenge Our Assumptions

Scenarios challenge our assumptions about the future. They focus on irreducible uncertainty: future changes that diverge from past observations. Based on the interactions among variable trajectories of multiple drivers, scenarios explore hypothetical boundary conditions beyond the scope of predictive models' assumptions. Scenarios become powerful when combined with predictive modeling. They are not an alternative to models but a complement to them, expanding boundary conditions and linking multiple social and ecological models in an integrated framework. Using the expanded boundary conditions set by the divergent scenarios, integrated models can help us to accomplish three tasks: (1) test hypothesized trajectories and interactions; (2) refine potential relationships and feedback among variables; and (3) assess potential impacts that hypothesized futures will have on ecosystem services and human well-being. Scenario planning provides a systematic approach to dealing with uncertainties in assessing alternative strategic actions.

Highlight Risks and Opportunities

A fundamental objective of scenario planning is to explore the interactions between multiple critical uncertainties, thus entertaining potential future conditions that might otherwise be overlooked. Scenarios attempt to highlight risks and opportunities of plausible future conditions by looking at divergent trajectories. If planners and decision-makers consider multiple divergent scenarios, they can engage in a creative process for imagining solutions.

Provide Warning Signals

Scenarios also help decision-makers to anticipate potential regime shifts by helping to reveal warning signals that may allow them to change their strategies in a timely and effective way. Robust strategies are effective under divergent futures, but adaptive strategies support effective action under specific conditions. Critical sensitivities refer to potential thresholds or irreversible conditions with significant implications for multiple ecosystem services and diverse stakeholders.

Make Robust Decisions

Scenarios help decision-makers to identify robust decisions and then to adopt and act on those decisions, despite uncertainty, by providing a systematic way to assess the robustness of alternative strategies under a set of plausible future conditions. The divergent future conditions that could emerge from the interaction of uncertain trajectories might help us to see the different effects of a major climate change versus a minor one in terms of magnitude and variability. They could also help us to recognize diverse trajectories of change in social values that characterize relationships between society and nature.

10

Building Cities That Think like Planets

IF OUR CITIES ARE TO BE RESILIENT ON A PLANETARY TIME SCALE, WE must expand our horizons of time and space as well as our ability to embrace change. Earth has evolved into a living planet over a billion years, allowing human life to emerge. How can thinking on a planetary scale help us understand the place of humans in the evolution of Earth and guide us in building a human habitat of the "long now"? This chapter discusses the implications of complexity and uncertainty for building the cities of the future, and it articulates pathways and principles for urban design and planning.

Cities face an important challenge: they must rethink themselves in the context of planetary change. Urban ecologists must understand the role that cities play in the evolution of Earth. Can the emergence and rapid expansion of cities across the globe represent a turning point in the life of our planet on a scale similar to that of the Great Oxidation (Lenton and Williams 2013)? And can the patterns of urbanization determine the probability of crossing thresholds that will trigger a planetary shift of the same magnitude and significance? I do not answer these questions. Only a new extraordinary collaboration among urban ecologists, evolutionary biologists, and other natural and social scientists might do so. I suggest that we begin to define a series of hypotheses and develop long-term studies that can tackle such questions in new and productive ways. We must also rethink the research infrastructure necessary to support diverse networks of scientific teams.

In this book, I have advanced the hypothesis that cities are hybrid ecosystems: they are unstable and at the same time able to change and innovate. I proposed a co-evolutionary paradigm for building a science of cities that "think like planets," a view that focuses on both unpredictable dynamics and experimental learning. I have elaborated on some concepts and principles of design and planning that emerge from such a perspective: self-organization, heterogeneity, modularity, cross-scale interactions, feedbacks, and transformation. In closing, I pose a question: how can thinking on a planetary scale help us to understand the place of humans in the evolution of Earth and guide us in building a human habitat of the long now?

Planetary Scales

Humans make decisions at multiple scales of time and space simultaneously, depending on how they perceive the scale of a given problem and the scale of influence that their decisions might have. Yet it is unlikely that the scale extends beyond one generation or includes the entire globe. The human experience of space and time has profound implications for our understanding of world phenomena and for making long- and short-term decisions. In his book *What Time Is This Place?*, Kevin Lynch (1972) eloquently told us that time is embedded in the physical world that we inhabit and build. Cities reflect our experience of time, and the way we experience time affects the way we view and change the environment. Thus our experience of time plays a crucial role in whether we succeed in managing environmental change. If we are to think like a planet, we must deal with scales and events that are far removed from everyday human experience. Earth is 4.6 billion years old. That's a big number to even conceptualize, much less incorporate meaningfully into our individual and collective decisions.

Thinking like a planet implies expanding the temporal and spatial scales of city design and planning, but not simply from local to global and from a few decades to a few centuries. Instead we must include a broad range of scales, from the human experience of place (Beatley 2010) and the landscape ecology of regions (Forman 2008) to the scale of geological and biological processes operating on the planet (Alberti 2014). Thinking on a planetary scale also requires expanding the idea of change. Lynch (1972: 1) reminded us that "the arguments of planning all come down to the management of change."

But what is change? Human experience of change is often confined to fluctuations within a relatively stable domain. However, Planet Earth has displayed rare but abrupt changes and regime shifts in the past (e.g., the last glacial-interglacial transition). Human experience of such change is limited to marked changes in regional system dynamics, such as altered fire regimes and extinctions of species. Yet since the Industrial Revolution, humans have been pushing the planet outside the domain of stability that it has occupied since the beginning of human history. Will human activities trigger a dramatic global event comparable to those experiences by Planet Earth? We can't answer that, as we don't understand enough about how regime shifts propagate across scales, but emerging evidence does suggest that if we continue to disrupt ecosystems and the climate, we face an increasing risk of crossing the thresholds that keep the Earth in a relatively stable domain. Until recently, our individual behaviors and collective institutions have been shaped primarily by change that we can envision relatively easily on a human time scale. Our behaviors are not tuned to the slow and imperceptible—but systematic— changes that can drive dramatic shifts in Earth's systems.

Planetary shifts can be rapid: the glaciation of the Younger Dryas, an abrupt climatic change resulting in severe cold and drought, occurred roughly 11,500 years ago, apparently over only a few decades. Or such shifts can unfold slowly: the Himalayas took over a million years to form. Shifts can emerge as the result of extreme events such as volcanic eruptions, or of relatively slow processes, such as the movement of tectonic plates. Though we still don't completely understand the subtle relationship between local and global stability in complex systems, several scientists hypothesize that the increasing complexity and interdependence of socioeconomic networks can produce "tipping cascades" and "domino dynamics" in the Earth's system, leading to unexpected regime shifts (Helbing 2013; Hughes et al. 2013).

Planetary Challenges and Opportunities

A planetary perspective on envisioning and building cities that we would like to live in—cities that are livable, resilient, and exciting—provides many challenges and opportunities. To begin, it requires that we expand the spectrum of imaginary archetypes. Current archetypes, from biological determinism to an equally narrow techno-scientific optimism, reflect skewed and often extreme simplifications of how the universe works (figure 10.1). At best, they are accurate but incomplete accounts of how the world works. How can we reconcile the messages contained in catastrophic-versus-optimistic views of the future of Earth? And how can we hold divergent explanations and arguments as plausibly true? Can we imagine a place where humans have co-evolved with natural systems? What does that world look like? How can we create that future in the face of limited knowledge and uncertainty, holding all these possible futures as plausible options?

The concept of planetary boundaries offers a framework for humanity to operate safely on a planetary scale. Rockström et al. (2009) developed the concept to inform us about the levels of anthropogenic change that can be sustained while avoiding potential planetary regime shifts that would dramatically affect human well-being. The concept does not imply, nor does it exclude, planetary-scale tipping points associated with human drivers. Hughes et al. (2013) addressed some misconceptions surrounding planetary-scale tipping points, especially the confusion of a system's rate of change with the presence or absence of a tipping point. To avoid the potential consequences of unpredictable planetary-scale regime shifts, we must shift our attention toward drivers and feedbacks rather than focusing exclusively on the detectable system responses. Rockström et al. (2009) also identified nine areas that are most in need of set planetary boundaries: climate change, biodiversity loss, input of nitrogen and phosphorus into soils and waters, stratospheric ozone depletion, ocean acidification, global consumption of freshwater, changes in land use, air pollution, and chemical pollution.

To provide a planetary perspective at the scale where policy development and decision-making occur, Steffen et al. (2015) further advanced the planetary boundary concept by introducing a two-tier approach to emphasizing cross-scale interactions and accounting for regional-level heterogeneity. They identified two key boundaries—climate change and biosphere integrity—as system-scale emergent properties of a tightly connected web of biophysical processes that have the potential to drive the Earth system into a new state. However, they did not account for either the regional distribution of the impact or historical patterns— important dimensions that, they acknowledged, are essential if the

FIGURE 10.1   Archetypes. Sources (*clockwise from top left*): Dutch Delta Works;
East River Blueway Plan/WXY Studio; Aqualt-by-Studio-Lindfors; Andrew Burton,
Getty Images.

framework is to serve as a guide for human development toward global sustainability.

A different emphasis is proposed by those scientists who have advanced the concept of planetary opportunities: solution-oriented research to provide realistic, context-specific pathways to a sustainable future (DeFries et al. 2012). The idea is that we must shift our attention to how human ingenuity and creativity can expand our ability to enhance human well-being (through, e.g., improved food security and human health) while minimizing and reversing environmental impacts. The concept is grounded in human innovation and our capacity to develop alternative technologies, implement green infrastructure, and reconfigure institutional frameworks. Opportunities to develop solution-oriented research approaches and explore innovative policy strategies are amplified on an urbanizing planet, where such solutions can be replicated and can transform the way we build cities and inhabit the Earth.

Imagining a Resilient Urban Planet

While the different images of the future in the emerging archetypes are both plausible and informative, they speak about the present more than the future. They are all extensions of the current trajectory, as if the future would unfold along a path defined by our current ways of asking questions and our current ways of understanding and solving problems. Yes, these perspectives do account for uncertainty, but it is defined by the confidence intervals around an extrapolated trajectory. And both stories are grounded in the inevitable dichotomies of humans versus nature and technology versus ecology. These views are, at best, an incomplete account of what is possible: they reflect a limited ability to imagine the future beyond such archetypes. Why can we imagine smart technologies but not smart behaviors, smart institutions, and smart societies? Why think only of technology and not of humans and their societies, which co-evolve with other life on Earth?

Understanding the co-evolution of human and natural systems is key to building a resilient society and transforming our habitat. Among the greatest questions in biology today is whether natural selection is the only process driving evolution and, if not, what the other potential forces might be. To understand how evolution constructs the mechanisms of life, molecular biologists would argue that we must first understand the self-organization of genes governing the evolution of cellular processes and influencing evolutionary change (Johnson and Kwan Lam 2010). In order to adequately understand the co-evolution of human and natural systems, we must embrace a complementary perspective on the forces driving evolution and the role that natural selection and self-organization play in constraining natural selection at different stages of the evolutionary process.

To function, life on Earth depends on the close cooperation of multiple elements. Biologists are curious about the properties of complex networks that supply resources, process waste, and regulate the system's functioning at various scales of biological organization. West and Brown (2005) proposed that natural selection solved this problem by evolving hierarchical, fractal-like branching. Other characteristics of evolvable systems include flexibility (e.g., phenotypic plasticity) and novelty. This capacity for innovation is an essential precondition for any system to function. Gunderson and Holling (2002) have noted that if systems lack the capacity for innovation and novelty, they may become overconnected and dynamically locked, unable to adapt. To be resilient and evolve, they must create new structures and undergo dynamic change. Differentiation, modularity, and cross-scale interactions of organizational structures have been described as key characteristics of systems that are capable of simultaneously adapting and innovating (Allen and Holling 2010).

To understand the co-evolution of human-natural systems also requires advances in the social theories that explain how complex societies and cooperation have evolved. What role does human ingenuity play? In this book, I have proposed that coupled human-natural systems are not governed only by either natural selection or human ingenuity but by hybrid processes and mechanisms. It is their hybrid nature that makes them unstable and, at the same time, able to innovate. This novelty of hybrid systems is key to reorganization and renewal. Urbanization modifies the spatial and temporal variability of resources, creates new disturbances, and generates novel competitive interactions among species. This is particularly important because the distribution of ecological functions within and across scales is key to a system's ability to regenerate and renew itself (Peterson, Allen, and Holling 1998).

Will the human species rise to the challenge and opportunity that nature has given us? Can we evolve in cooperation with other species toward a hybrid planet (Frank, Alberti, and Kleidon 2016)? Scientists around the world have alerted us to our role as a geological force that shapes the global landscape and evolution of Planet Earth (Crutzen and Stoermer 2000). We are driving the sixth mass extinction (Kolbert 2015) and changing our climate (IPCC 2014). Turning the tide will require a significant shift in the human enterprise that only the power of people can generate through their collective imagination and action. The climate deal signed by 195 nations on December 13, 2015 (UNFCCC 2015), might signal humanity's emerging awareness of our responsibility to the planet. Whether we are actually able to keep the global temperature within 2 degrees Celsius of preindustrial levels, as agreed in this deal, has yet to be seen. But if we are to be successful, it is cities that must lead in this ambitious transformation.

The City That Thinks Like a Planet: What Does It Look Like?

Although I have ventured to pose this question, I will not attempt to provide an answer. In fact, no single individual can. The answer resides in the collective imagination and evolving behaviors of peoples of diverse cultures who inhabit the vast array of regions across the planet. Humanity has the capacity to think in the long term. Indeed, throughout history, people in societies faced with the prospect of deforestation or other environmental changes have successfully engaged in long-term thinking. Consider, for example the Tokugawa shoguns, Inca emperors, New Guinea highlanders, and sixteenth-century German landowners (as discussed in Diamond 2005) or, more recently, the Chinese efforts at reforestation and bans on logging of native forests.

Many European countries, and the United States, have dramatically reduced their air pollution while increasing their use of energy and their combustion of fossil fuels. Humans have the intellectual and moral capacity to do even more when they tune in to challenging problems and engage in solving them.

A city that thinks like a planet is not built on previously set design solutions or planning strategies. Nor can we assume that the best solution would work equally well across the world, regardless of place and time. Instead, such a city must be built on principles that expand its drawing board and collaborative actions that include planetary processes and scales to integrate humanity into the evolution of Earth. Such a view acknowledges the history of the planet in every element or building block of the urban fabric, from the building to the sidewalk, from the backyard to the park, from the residential street to the highway. It is a view that is curious about understanding who we are and about taking advantage of novel patterns, processes, and feedbacks that emerge from human and natural interactions. It is a city grounded in the here and the now and simultaneously in the different temporal and spatial scales of human and natural processes that govern the Earth. A city that thinks like a planet is simultaneously resilient and able to change.

How can such a perspective guide decision-making in practice? Urban planners and other decision-makers strategizing about investments in public infrastructure want to know whether there are generic properties of city structure and governance that predict the capacity to adapt and transform. Can such shifts in perspective, toward a planetary view, provide a new lens and interpretation of the evolution of human settlements, adaptation to change, and success? Emerging evidence from the study of complex systems indicates that systems with greater heterogeneity (e.g., economic systems with diverse economic activities and ecosystems with a diversity of species) and greater modularity (where some components are independent) tend to have greater adaptive capacity than those characterized by highly connected, homogeneous elements (Scheffer et al. 2012).

Other properties of complex systems that enhance adaptive capacity and allow innovation are cross-scale interaction, early warning, and self-organization (figure 10.2). How do these qualities apply to cities?

Diversity allows systems to be flexible and cities to function under a wide range of conditions (e.g., with multiple modes of transportation). In October 2012, after Hurricane Sandy, the subway in New York City flooded unexpectedly and shut down for a week, potentially disrupting many interconnected activities. Yet it is under such circumstances that we encounter the greatest surprises: in New York, the flexibility created by an imperfect and redundant urban infrastructure provided alternatives and workarounds.

Modularity facilitates autonomous functionality and allows a system such as the urban built infrastructure to contain disturbances and avoid cascading effects. The modular electric grid is a prime example: interdependent networks are arranged such that a failure in one place leads to failures in other places. Connectivity may stabilize some processes, by reinforcing generator pathways, but it may also create unstable power systems that propagate failure across networks that are highly interdependent. Modular electric grids make it possible to isolate parts of the systems, which may then continue to operate independently and provide substitutive functionality during extreme events.

Cross-scale interactions allow for functional redundancy across scales and allow a system to respond to disturbance by exploiting innovative options for service substitutions (e.g., energy, food, and water sources and delivery). Like cross-scale resilience, which is produced by the species of a functional group that operate across scales, urban infrastructure sys-

Resilience Principle	Hypotheses	Example
Heterogeneity	Allows system flexibility and the ability to function under a wide range of conditions	Multimodal transportation 
Modularity	Allows autonomous functionality and the ability to contain disturbances and avoid cascading effects	Modular electric grid 
Cross-scale interactions	Allows functional redundancy across scales, added capacity under contingency, and creative solutions for service substitutions	Energy and water sources and delivery 
Early warning	Anticipating catastrophic events and allowing the system to fail safely also depends on creating early warning systems that allow for essential functions to be performed when part of the system fails	Realtime monitoring systems 
Self-organization	Resilient systems are also self-organizing, a quality that enables natural and social systems to change their internal structure and their function in response to external circumstances	Sand ripples and stock markets 

FIGURE 10.2   Resilience principles: emerging hypotheses regarding properties of resilient complex systems. Photographs: bike on bus: SounderBruce; electric grid: Chad Cooper; water delivery: Layne Construction pipeline (Pueblo, Colorado); warning signals: U.S. Army; sand dune: ccdoh1.

tems that operate multiple services reinforce system functionality and thus enhance resilience.

Early warning is an essential component of adaptation and transformation. Anticipating catastrophic events by establishing early warning systems, together with a modular architecture, allows a system to fail safely by permitting essential functions to be performed while part of the system fails. Early warning mechanisms (e.g., indicators, real-time warnings, and crash zones) provide real-time feedback and help us to anticipate catastrophic events.

Self-organization is a process in which patterns at the global level of a system emerge from numerous interactions among the lower-level components of the system itself. It enables natural and social systems to change their internal structures and their functions in response to external circumstances (e.g., sand ripples or stock market shifts). Novel forms of social organizations and socioecological functions can emerge from interactions at lower scales among multiple agents.

Translating These Concepts into Practice

A co-evolutionary perspective shifts the focus of planning toward human-natural interactions, adaptive feedback mechanisms, and flexible institutional settings. Instead of predefining "solutions" that communities must implement, such a perspective focuses on understanding the rules of the game in order to facilitate self-organization and carefully balanced top-down and bottom-up management strategies. Planning relies on principles that expand the heterogeneity of forms and functions in the structures and infrastructures that support a city, and plans support selective modularity (as opposed to generalized connectivity) to create interdependent decentralized systems with adequate autonomy to evolve.

Local governments must make important decisions about land-use management and investments in infrastructure that can influence the direction of urban development in the near future. Future policies and management practices will succeed or fail based on their ability to take into account the complexities and uncertainties of these systems. To address the inherent uncertainty of coupled human-natural systems, we must challenge the myths of stability and optimality that have long characterized urban planning. The search for optimal solutions is based on three assumptions: that thresholds remain constant over time, that they can be detected and predicted, and that what is resilient in one region and at one scale is resilient across regions and across scales. Yet increasing evidence challenges the universality of these assumptions.

Resilience depends on biophysical and socioeconomic conditions that vary across regions and scales. The most recent IPCC report (IPCC 2014) illustrates significant differences in both the potential risks that urban communities across different regions will face and their adaptive capacities. According to the IPCC Working Group II, which focuses on urban areas, four factors explain differences across cities: local government capacity, the proportion of residents served with risk-reducing infrastructure and services, the proportion living in housing built to appropriate health and safety standards, and the levels of risk from climate change's direct and indirect impacts (Revi et al. 2014). The report does caution that the current evidence is still too limited to draw conclusions that explain adaptive capacity and resilience across communities, but it is increasingly clear that cities will require different, location-specific strategies in order to adapt to environmental change and build resilience (ibid.).

I suggest five principles for planning under uncertainty that can enable resilience and innovation in urban ecosystems:

1. Maintain the diversity of urban patterns and processes and a variety of infrastructure typologies to support diverse human and ecosystem functions rather than aim for an optimal city design.
2. Focus on maintaining self-organization and increasing adaptation capacity rather than aiming to reduce variability, control change, and eliminate uncertainty.
3. Challenge assumptions of current policies and strategies and actively reconfigure problem definitions and policy actions.
4. Create options for learning through experiments and design flexible strategies that allow us to adapt them and implement what we learn.
5. Expand the time scale of decision-making by designing policies and strategies to be robust under divergent but plausible futures.

In cities across the world, people are setting examples that will allow these principles to be tested. Human perceptions of time and experiences of change are emerging as keys to achieving the shift to a new perspective for building cities. We must develop reverse experiments to explore what works—what shifts the time scales of individual and collective behaviors. We must direct our attention toward the opportunities that urban transformations may provide for an urban planet. Which lessons for urban ecology can we draw from the evolution of urban systems and their persistence through time?

Translating principles of resilience and innovation to the practice of urban design and planning requires a shift from our current fragmented search for solutions toward an integrated and dynamic view of urban systems. The form and infrastructure of resilient hybrid cities are heterogeneous and modular. Hybrid cities rely on innovative technology and self-organized governance. Transportation, energy, water, and waste management systems are interconnected and link housing, employment, and business clusters across multiple scales. Local governments of these cities invest in innovative infrastructure and technologies (e.g., bus rapid transit, or BRT) and in new forms of finance (e.g., Qualified Energy Conservation Bonds, or QECBs) and governance (e.g., horizontal governance).

Emerging evidence suggests that the most innovative cities are typically more productive, more socially inclusive, and more resilient and that they have lower carbon emissions (Floater et al. 2014). Many cities across the world are implementing innovative solutions to foster new urban development models and accelerate transformation. New technologies and the digital revolution have provided unanticipated opportunities for increasing efficiency by sharing resources, managing service demands, monitoring, and providing urban dwellers the real-time, people-centered information required to make informed decisions (World Economic Forum 2015). Smart transportation systems—including, for example, BRT, bicycle "superhighways," car and bicycle sharing, smarter traffic information systems, and electric vehicles—have appeared in the past decade in cities across Europe, the Americas, China, and India (figure 10.3).

Several Northern European cities have adopted successful strategies to cut greenhouse gases, combining these strategies with innovative approaches that allow the cities to adapt to the inevitable consequences of climate change. One example is the Copenhagen 2025 Climate Plan, which lays out a path for Copenhagen to become the world's first carbon-neutral city by 2025 through efficient zero-carbon mobility and building. The city is building a subway project that will place 85 percent of its inhabitants within 650 yards of a metro station. Nearly three-quarters of Copenhagen's emissions reductions will be realized as people transition to less carbon-intensive ways to produce heat and electricity via a diverse supply of clean energy: biomass, wind, geothermal, and solar. Copenhagen is also one of the first cities to adopt a climate adaptation plan that will reduce its vulnerability to the extreme storm events and rising seas expected over the next hundred years.

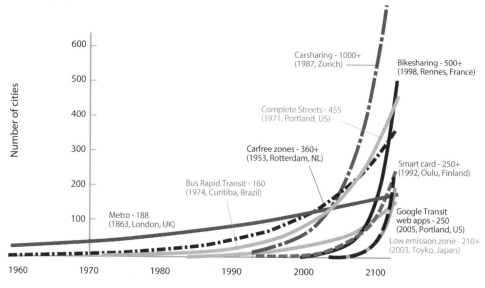

FIGURE 10.3   Smart transportation systems in selected cities. Source: Embarq 2013.

In the Netherlands, alternative strategies are being explored to allow people to live with the inevitable floods. These strategies include building floating communities on water and engineering and implementing adaptive beach protections that take advantage of natural processes. The Sand Engine, completed in 2011, is an experimental project developed by Building with Nature (a consortium of Dutch industries, universities, research institutes, and public water agencies) that uses a combination of wind, waves, tides, and sand to replenish the eroded coasts. The Dutch Rijkswaterstaat, the executive agency of the Dutch Ministry of Infrastructure and the Environment, and the South Holland provincial authority deposited enormous quantities of sand as a one-by-two-kilometer artificial peninsula extending into the sea; waves and currents will redistribute it over time, building dunes and beaches to protect the coastline.

New York is setting an example for long-term planning by combining adaptation and transformation strategies into its plan to build a resilient city: in the summer of 2013, in the wake of Hurricane Sandy, then-mayor Michael Bloomberg outlined a US$19.5 billion plan to defend the city against rising seas. In many rapidly growing cities across the world, similar leadership is emerging. For example, in 2009, Johannesburg adopted one of the first climate change adaptation plans, as have Durban and Cape Town in South Africa and Quito in Ecuador, along with Ho Chi Minh City in Vietnam, which has established a partnership with Rotterdam in the Netherlands to develop a resilience strategy.

These new approaches to smart design and technologies are supported by novel financial mechanisms. Cities are using municipal bonds to finance innovative infrastructure projects that can attract large investors. In 2013, Johannesburg issued a US$136 million green bond to finance investments in a diverse set of solutions, from hybrid buses to biogas energy and rooftop solar water heaters. In the United States, models such as QECBs allow local governments to borrow money to fund energy conservation projects. Municipal bonds and infrastructure trusts have significant potential to accelerate transformation.

Novel institutional frameworks and policy approaches that embrace uncertainty and build resilience are being tried experimentally in a few metropolitan areas. Wise et al. (2014) examined how institutional settings and governance may facilitate or constrain adaptation. Rather than focusing on adaptation as an isolated set of interventions, they conceptualized adaptation as a continual pathway of change and response informed by past actions and shaping the range of future options. This approach, called *adaptation pathways*, emphasizes five critical elements: multidimensionality, interdependence, intertemporality, monitoring, and social processes.

Multidimensionality acknowledges that climate adaptation cannot be separated from the cultural, political, economic, environmental, and developmental contexts in which it occurs; it must be thought of as part of a range of societal responses to change. *Interdependence* recognizes interactions across spatial scales, sectors, and jurisdictional boundaries, which can lead to threshold effects and require coordination. *Intertemporality* implies that future pathways are contingent on historical pathways; it acknowledges the potential for lock-in and warns us to be wary. *Monitoring* means the implementation of early warning systems and detection of warning signals to assess dynamic changes and the effectiveness of interventions and to respond adaptively. *Social processes* enable or constrain the rules, values, and knowledge that govern adaptation and resilience and that must be embedded in planning strategies

**Box 10.1. Pathways to Resilience:
The New York City Example**

In 2008, Mayor Michael Bloomberg convened the First New York City Panel on Climate Change (NPCC 2010) to assist the city in assessing risk and developing an adaptation strategy. A major innovative element introduced by the panel was the concept of "flexible adaptation pathways" as an approach to responding to climate change (Yohe and Leichenko 2010). New York's flexible adaptation approach combines two major concepts: *resilience* and *dynamic robustness* (W. E. Walker, Haasnoot, and Kwakkel 2013). This approach is based on a study done by the City of London and the UK Environment Agency to renovate the Thames barriers (Lowe et al. 2008; Ranger, Reeder, and Lowe 2013).

The level of risk that is acceptable to society fluctuates as a result of changes (e.g., an experience of a major hurricane). Without climate change mitigation or adaptation, inflexible adaptation standards may improve conditions on a relatively short time line but eventually will result in crossing the acceptable risk level. Flexible adaptation consists

to ensure that adaptive pathways succeed. Rosenzweig and Solecki (2014) provided a practical example of this approach and showed how it might be applied in their assessment of the flexible adaptation pathways being developed to manage climate risk in response to Hurricane Sandy (box 10.1).

To think like a planet and explore what is possible, we may need to reframe our questions. Instead of asking what is good for the planet, we must ask what is good for a planet inhabited by people, or what a good human habitat on Earth might be. And instead of seeking optimal solutions, we should identify principles that will inform diverse communities across the world. The best choices may be temporary, since we do not fully understand the mechanisms of life and we cannot predict the consequences of human actions. They may very well vary with place and depend on their own histories, but human actions may also constrain the choices available for life on Earth.

of a successive set of strategies developed and implemented as knowledge and understanding of climate change proceed. These pathways are reevaluated and readjusted over time as new knowledge and information become available.

The city is developing short-term actions that reduce risk iteratively while laying a framework to guide long-term flexibility and resilience. Proposed adaptation policies for the city include reducing flood risks to infrastructure, buildings, and highly exposed communities through small- to medium-scale flood protection strategies (e.g., levees), regulatory approaches (e.g., building codes), and improving responses to extreme events (New York City 2011). Mitigation actions may be necessary for the adaptation to succeed. New York's flexible adaptation framework encompasses both mitigation and adaptation and enables planners to consider long-range goals and ways to translate them into short-term objectives. Based on an initial report by the NPCC, the city updated PlaNYC, its long-term sustainability plan, by explicitly acknowledging that risk management strategies must evolve through time in response to continuous climate risk assessment, evaluation of adaptation strategies, and monitoring. •

Scenario Planning

Scenario planning offers a systematic and creative approach to thinking about the future. It challenges scientists and practitioners to expand their mindsets (see chapter 9). It provides a tool that we can use to deal with the limited predictability of changes on the planetary scale and to support decision-making under uncertainty. Scenarios help bring the future into present decision-making processes (Schwartz 2005), broaden perspectives, prompt new questions, and expose possibilities for surprise. They have several other valuable features: they can shift attention toward resilience, redefine decision frameworks, expand the boundaries of predictive models, highlight the risks and opportunities of alternative future conditions, monitor early warning signals, and identify robust strategies (Alberti, Russo, and Tenneson 2013).

A fundamental objective of scenario planning is to explore interactions among uncertain trajectories that would otherwise be overlooked.

Scenarios highlight the risks and opportunities of plausible future conditions. The hypothesis is that if planners and decision-makers look at multiple divergent scenarios, they will engage in a more creative process for imagining solutions that would be otherwise invisible. As I noted earlier, scenarios are narratives of plausible futures; they are not predictions. But they are extremely powerful when combined with predictive modeling. They help expand boundary conditions and provide a systematic approach that we can use to deal with intractable uncertainties and to assess alternative strategic actions, and they can help us modify model assumptions and assess the sensitivities of model outcomes. Building scenarios can help us highlight gaps in our knowledge and identify the data required to assess future trajectories.

Scenarios can also highlight important warning signals, allowing decision-makers to anticipate unexpected regime shifts and respond in timely and effective ways. They can support decision-making in uncertain conditions by providing a systematic way to assess the robustness of alternative strategies under a set of plausible future conditions. Although we do not know the probable impacts of uncertain futures, scenarios provide the basis for assessing critical sensitivities and help us to identify both potential thresholds and irreversible impacts, allowing us to maximize the well-being of both humans and our environment.

A New Ethic for a Hybrid Planet

More than half a century ago, Aldo Leopold (1949) introduced the concept of "thinking like a mountain": he wanted to expand the spatial and temporal scale of land conservation by incorporating the dynamics of the mountain. Defining a land ethic, as Leopold did, was a first step in acknowledging that we are all part of a larger community that includes soils, waters, plants, and animals and all the components and processes that govern the land, including the prey and predators. Along the same lines, Paul Hirsch and Bryan Norton (2012) have articulated a new environmental ethic by suggesting that we "think like a planet." Building on their idea, I propose that we must expand the dimensional space of our mental models of urban design and planning to the planetary scale.

Glossary

This glossary is based in part on definitions compiled in Alberti 2008, Alberti 2015, and Resilience Alliance 2015.

adaptive capacity: A system's ability to reconfigure itself in the face of change without significant declines in crucial functions such as primary productivity, hydrological cycles, social relations, and economic prosperity (Folke et al. 2002).

adaptive cycle: A metaphor used to describe four commonly occurring phases of change in complex systems: exploitation, conservation, creative destruction, and renewal (also referred to as r, K, omega, and alpha) (Holling 1986).

agent: An autonomous decision-maker. The word agent is derived from the Greek agein, which means "to drive or lead," and from Latin agere, which means "to act."

Anthropocene: A proposed geologic and chronological term for the epoch beginning when human activities started to exert a significant global impact on the Earth's ecosystems (Crutzen 2002).

biodiversity: A contraction of *biological diversity*. The term refers to the number, variety, and variability of the living organisms in a physical or temporal space; it may also refer to diversity within a species (genetic diversity), between species (species diversity), or between ecosystems (ecosystem diversity) (MEA 2005).

biotic homogenization: The process by which species invasions and extinctions increase the genetic, taxonomic, or functional similarity of two or more locations over a specified interval of time (Olden 2008).

carbon cycle: The biogeochemical cycle in which carbon is exchanged among Earth's biosphere, pedosphere, geosphere, hydrosphere, and atmosphere (Falkowski et al. 2000).

complex system: A system with properties not fully explained by an understanding of its component parts. Its behavior is nonlinear, and it exhibits structural and functional characteristics that emerge from the interactions of its constituent parts (Goldenfeld and Kadanoff 1999).

coupled human and natural system: A system in which human and natural components interact (Liu et al. 2007a).

critical state, critical point: In thermodynamics, the endpoint of a phase equilibrium curve.

cross-scale interactions: The effects exerted, or experienced, by the dynamics of a given system at a given scale on, or due to, the dynamics of the scales that are embedded within that system or that enfold it (Holling, Gunderson, and Peterson 2002).

discontinuity hypothesis: Derived from hierarchy theory, this hypothesis predicts weaker interactions among species operating across markedly different scales compared to those among species operating at similar scales (Nash et al. 2014).

disturbance: A discrete event in time that causes a temporary change in average environmental conditions. A disturbance may disrupt ecosystems, communities, or population structures, and it may impact resources, substrate availability, or the physical environment (White and Pickett 1985).

eco-evolutionary feedbacks: Reciprocal interactions between ecological and evolutionary dynamics on contemporary time scales (Palkovacs and Hendry 2010).

ecological niche: The functional role and position of a species in the ecosystem, including the resources it uses and its interactions with other species (Schoener 2009).

ecosystem: A subset of the organisms, environments, and interactions among the organisms and environments in a given area that, through flows of energy, lead to clearly defined trophic structures, biotic diversity, and material cycles (e.g., the exchange of materials between living and nonliving parts) (Odum 1971).

ecosystem function: The flux of energy, organic matter, or nutrients in an ecosystem, including the flux of biomass associated with trophic interactions. Function is expressed as a rate of change of an ecosystem property (Likens 1992).

emergent property: A phenomenon that is not evident in the constituent parts of a system but which appears when those components interact within the overall system (O'Connor and Wong 2015).

Great Oxidation Event (GOE): The biologically induced appearance of dioxygen (O_2) in Earth's atmosphere that occurred about 2.3 billion years ago (2.3 Ga). Cyanobacteria, which appeared about 200 million years before the GOE, began producing oxygen by photosynthesis, and the excess free oxygen began to accumulate in the atmosphere, causing the extinction of anaerobic organisms. Eventually aerobic organisms began to evolve, consuming oxygen and bringing oxygen availability into equilibrium (Sosa Torres, Saucedo-Vázquez, and Kroneck 2015).

hierarchy theory: An evolution of general systems theory that has emerged as part of the general science of complexity. Hierarchy theory focuses on levels of organization and scale. These nested levels emerge according to the dominant spatiotemporal scales at which system elements operate, causing them to interact most strongly with other elements of the same type, less strongly with elements that are dissimilar, and more weakly with patterns, processes, and elements operating at disparate scales. Hierarchies occur in both social and ecological systems (Ahl and Allen 1996).

human function: Socioeconomic, political, and residential/interpersonal processes, institutions, and interactions that sustain and maintain human populations (Alberti and Marzluff 2004).

hybrid ecosystem: In ecology, the transitional state of an ecosystem between historical and novel states (Hobbs, Higgs, and Harris 2009). The hybrid state is a new state able to return to its historical functions, while a novel ecosystem is characterized by irreversible regime shifts. This transitional state has previously been conceptualized as an area on a two-dimensional system defined by axes representing the abiotic and biotic distances (or both) from the historical condition. I propose that to begin to understand the dynamic of novel coupled human-natural systems, we must include a third axis: evolution of the human habitat from Neolithic settlements to industrial and postindustrial cities.

hypervolume: Defines the multidimensional space of resources (e.g., light, nutrients, structure, etc.) available to, and specifically used by, an organism. The Hutchinsonian niche, for example, is an n-dimensional hypervolume, in which the dimensions are the environmental conditions and resources that an individual or species requires to practice its way of life (i.e., to persist) (Hutchinson 1957).

inverse problem: A problem solved by identifying, calculating, or estimating the causative factor(s) responsible for a given (set of) observable outcome(s) (Kabanikhin 2008).

landscape: An area that is spatially heterogeneous in at least one factor of interest (M. G. Turner, Gardner, and O'Neill 2001).

long now: Brian Eno coined the term "long now" in a 1978 speech, "The Big Here and Long Now," in New York City; http://longnow.org/essays/big-here-long-now.

natural habitat: The collection of biotic and abiotic elements within a given location—one predominantly unaltered by human activities—that provides the food and shelter a species needs to survive.

net primary productivity: A measure of the overall accumulation of energy in an ecosystem due to the conversion of solar energy into chemical energy that accounts for consumption (i.e., photosynthetic product minus respiration demands).

network theory: A theory that has evolved in mathematics and physics as part of graph theory; it has been applied in many fields, including computer science, biology, epidemiology, economics, and sociology. A network consists of a set of nodes joined by links. While initially represented by random graphs, real networks display some organizing principles (Alberti and Barabási 2002).

niche construction: The process by which organisms modify components of their environment, such as resource distribution or habitat space, to affect selection pressures on themselves or other organisms in an ecosystem (Odling-Smee 2003).

nutrient cycling: The processes by which elements are extracted from their mineral, aquatic, or atmospheric sources or recycled from their organic forms and then converted to ionic form as biotic uptake occurs, ultimately returning them to the atmosphere, water, or soil (MEA 2005).

power law: A functional relationship between two quantifiable entities in which the quantity of one entity varies as a power of the quantity of the other.

regime shift: A sudden shift in ecosystems whereby a threshold is passed and the core functions, structure, and processes of the new regime are fundamentally different from those of the previous regime (Scheffer and Carpenter 2003).

resilience: The capacity of a system to return to its original state following a disturbance. The size and shape of the basin of attraction around a stable system state define the maximum perturbation that the system can tolerate without shifting to an alternative stable state (Holling 1973).

scenarios: Plausible descriptions of how the future might develop, based on a coherent and internally consistent set of assumptions about the interactions of key uncertain and important driving forces, such as rate of technological innovation, economic development, or climate change (MEA 2005).

self-organization: A process in which some form of overall order or coordination emerges from local interactions between component parts of an initially disordered system. The process of self-organization is spontaneous, and it is not controlled by an agent outside the system. It can be triggered by random fluctuations that are amplified by positive feedback (Nicolis and Prigogine 1977).

slow versus fast variables: *Slow* variables are factors that change slowly in response to long-term processes and that constrain the responses of *fast* variables, leading to a potential shift in an ecosystem state (Chapin, Folke, and Kofinas 2009).

sprawl: Unplanned and uncontrolled growth of an urban area. Sprawl, which occurs at the margins of cities and towns, usually results in loss of rural areas and terrestrial habitats. It consumes land, its inhabitants are dependent on automobiles, and it typically exhibits population densities of one thousand to five thousand people per square kilometer (MEA 2005).

stability: The ability of a system to return to an equilibrium state after a temporary disturbance. The more rapidly it returns, and the less fluctuation it experiences, the more stable it is (Holling 1973).

tele-coupling: Emerging reciprocal interactions between distant natural and human systems (Liu et al. 2013).

threshold: A break point between two regimes of a system (B. Walker and Meyers 2004).

transformability: The capacity of a system to be fundamentally reformed in response to ecological, economic, or social (including political) conditions that render its existing form untenable (B. Walker et al. 2004).

urban: The U.S. Census defines a population agglomeration as *urban* if it has 2,500 or more inhabitants, generally with population densities of 1,000 or more persons per square mile. However, an urban area may be either densely populated (occupying a small land area) or distributed across a large land area of built infrastructure (with high overall population, high infrastructure density, and lower population density).

urban ecosystems: Coupled human-natural systems in which people are the dominant agents and are highly dependent on large inputs of materials and energy from beyond the system's boundaries and on vast outside capacities to absorb pollution and waste.

urban gradient: The continuum in the level of development (and associated patterns of land use and land cover) across a region, from the urban core(s) to rural fringe(s), and the associated continuum of differing human and biophysical processes (McDonnell and Pickett 1993).

urban system: A built environment with high human population density. Urban systems are operationally defined as human settlements with a minimum population density commonly in the range of four hundred to one thousand persons per square kilometer, a minimum population size typically between one thousand and five thousand people, and maximum agricultural employment usually in the vicinity of 50 to 75 percent (MEA 2005).

Bibliography

Ahl, V., and T. F. H. Allen. 1996. *Hierarchy Theory: A Vision, Vocabulary, and Epistemology*. New York: Columbia University Press.

Albert, R. 2006. "General Network Theory." In S. J. Hwang, W. J. Sullivan, and A. R. Lommel, eds., *LACUS Forum* 32: *Networks*, 3–20. Houston: Linguistic Association of Canada and the United States.

Albert, R., and A. L. Barabási. 2002. "Statistical Mechanics of Complex Networks." *Reviews of Modern Physics* 74, no. 1: 47.

Alberti, M. 2007. "Ecological Signatures: The Science of Sustainable Urban Forms." *Places* 19, no. 3: 56–60.

———. 2008. *Advances in Urban Ecology: Integrating Humans and Ecological Processes in Urban Ecosystems*. New York: Springer.

———. 2010. "Maintaining Ecological Integrity and Sustaining Ecosystem Function in Urban Areas." *Current Opinion in Environmental Sustainability* 2, no. 3: 178–84.

———. 2014. "Anthropocene City." In *Welcome to the Anthropocene: The Earth in Our Hands*, exh. cat., Deutsche Museum Special Exhibit 2014–2015, www.deutsches-museum.de/en/exhibitions/special-exhibitions/anthropocene/catalogue/.

———. 2015. "Eco-Evolutionary Dynamics in an Urbanizing Planet." *Trends in Ecology and Evolution* 30, no. 2: 114–26.

Alberti, M., and L. Hutyra. 2013. "Carbon Signatures of Development Patterns along a Gradient of Urbanization." In *Land Use and the Carbon Cycle: Advances in Integrated Science, Management and Policy*, edited by D. G. Brown, D. T. Robinson, N. H. H. French, and B. C. Reed, 305–28. Cambridge: Cambridge University Press.

Alberti, M., and J. Marzluff. 2004. "Ecological Resilience in Urban Ecosystems: Linking Urban Patterns to Human and Ecological Functions." *Urban Ecosystems* 7, no. 3: 241–65. doi:10.1023/B:UECO.0000044038.90173.c6.

———. 2013. Personal communication to author.

Alberti, M., J. M. Marzluff, E. Shulenberger, G. Bradley, C. Ryan, and C. Zumbrunnen. 2003. "Integrating Humans into Ecology: Opportunities and

Challenges for Studying Urban Ecosystems." *BioScience* 53: 1169.doi:10.1641
/0006-3568(2003)053[1169:IHIEOA]2.0.CO;2.

Alberti, M., M. Russo, and K. Tenneson. 2013. *Snohomish Basin Scenarios 2060*.
Seattle: Urban Ecology Research Lab, University of Washington. http://
urbaneco.washington.edu/wp/research/snohomish-basin-2060-scenarios/.
Accessed September 2015.

Alberti, M., and P. Waddell. 2000. "An Integrated Urban Development and
Ecological Simulation Model." *Integrated Assessment* 1, no. 3: 215–27.

Aligica, P. D. 2003. "Prediction, Explanation and the Epistemology of Future
Studies." *Futures* 35: 1027–40.

Allen, C. R., D. G. Angeler, A. S. Garmestani, L. H. Gunderson, and C. S.
Holling. 2014. "Panarchy: Theory and Application." *Nebraska Cooperative
Fish and Wildlife Research Unit—Staff Publications*. Paper 127. http://digital
commons.unl.edu/ncfwrustaff/127.

Allen, C. R., L. Gunderson, and A. R. Johnson. 2005. "The Use of Discontinuities and Functional Groups to Assess Relative Resilience in Complex
Systems." *Ecosystems* 8, no. 8: 958–66.

Allen, C. R., and C. S. Holling. 2008. *Discontinuities in Ecosystems and Other
Complex Systems*. New York: Columbia University Press.

———. 2010. "Novelty, Adaptive Capacity, and Resilience." *Ecology and Society*
15, no. 3: 24, www.ecologyandsociety.org/vol15/iss3/art24/.

Allen, P. M., and M. Sanglier. 1978. "Dynamic Models of Urban Growth." *Journal
of Social and Biological Structures* 1, no. 3: 265–80.

———. 1979. "A Dynamic Model of Urban Growth: II." *Journal of Social and
Biological Structures* 2, no. 4: 269–78.

Allen, T. F. H., and T. B. Starr. 1982. *Hierarchy: Perspectives for Ecological Complexity*. Chicago: University of Chicago Press.

Allison, S. D. 2004. "Microbial Allocation to Soil Enzyme Production: A Mechanism for Carbon and Nutrient Cycling." PhD diss., Stanford University.

Amanatidou, E., M. Butter, V. Carabias, T. Koennoelae, M. Leis, O. Saritas,
P. Schaper-Rinkel, and V. van Rij. 2012. "On Concepts and Methods in
Horizon Scanning: Lessons from Initiating Policy Dialogues on Emerging
Issues." *Science and Public Policy* 39: 208–21.

Anderies, J. M., S. R. Carpenter, W. Steffen, and J. Rockström. 2013. "The
Topology of Non-Linear Global Carbon Dynamics: From Tipping Points
to Planetary Boundaries." *Environmental Research Letters* 8, no. 4: 044048.
doi:10.1088/1748-9326/10/6/069501.

Anderies, J. M., M. A. Janssen, and E. Ostrom. 2004. "A Framework to Analyze
the Robustness of Social-Ecological Systems from an Institutional Perspective." *Ecology and Society* 9, no. 1: 18.

Anderson, C., K. Frenken, and A. Hellervik. 2006. "A Complex Network Approach
to Urban Growth." *Environment and Planning A* 38, no. 10: 1941–64.

Anderson-Teixeira, K. J., V. M. Savage, A. P. Allen, and J. F. Gillooly. 2009.
"Allometry and Metabolic Scaling in Ecology." In *Encyclopedia of Life
Sciences*, 1–9. New York: John Wiley and Sons.

Andersson, E., S. Barthel, S. Borgström, J. Colding, T. Elmqvist, C. Folke, and Å. Gren. 2014. "Reconnecting Cities to the Biosphere: Stewardship of Green Infrastructure and Urban Ecosystem Services." *Ambio: A Journal of the Human Environment* 43, no. 4: 445–53. doi:10.1007/s13280-014-0506-y.

Andrews, C. J. 2002. *Humble Analysis*. Westport, CT: Praeger.

Angel, S. 2012. *Planet of Cities*. Cambridge, MA: Lincoln Institute of Land Policy.

Angeler, D. G., S. Drakare, and R. K. Johnson. 2011. "Revealing the Organization of Complex Adaptive Systems through Multivariate Time Series Modeling." *Ecology and Society* 16, no. 3: 5.

Apicella, C. L., F. W. Marlowe, J. H. Fowler, and N. A. Christakis. 2012. "Social Networks and Cooperation in Hunter-Gatherers." *Nature* 481, no. 7382: 497–501.

Arbesman, S., J. M. Kleinberg, and S. H. Strogatz. 2009. "Superlinear Scaling for Innovation in Cities." *Physical Review E* 79, no. 1: 1–5. doi:org/10.1103/PhysRevE.79.016115.

Arnold, C. L., P. J. Boison, and P. C. Patton. 1982. "Sawmill Brook: An Example of Rapid Geomorphic Change Related to Urbanization." *Journal of Geology* 90, no. 2: 155–66.

Aronson, M. F. J., F. A. La Sorte, C. H. Nilon, M. Katti, M. A. Goddard, C. A. Lepczyk, P. S. Warren, et al. 2014. "A Global Analysis of the Impacts of Urbanization on Bird and Plant Diversity Reveals Key Anthropogenic Drivers." *Proceedings of the Royal Society B* 281: 20133330. doi:10.1098/rspb.2013.3330.

Arrhenius, O. 1921. "Species and Area." *British Ecological Society* 9, no. 1: 95–99.

Atwell, J. W., G. C. Cardoso, D. J. Whittaker, S. Campbell-Nelson, K. W. Robertson, and E. D. Ketterson. 2012. "Boldness Behavior and Stress Physiology in a Novel Urban Environment Suggest Rapid Correlated Evolutionary Adaptation." *Behavioral Ecology* 23, no. 5: 960–69.

Bak, P. 1996. *How Nature Works: The Science of Self-Organized Criticality*. New York: Springer.

Bak, P., C. Tang, and K. Wiesenfeld. 1987. "Self-Organized Criticality: An Explanation of the 1/f Noise." *Physical Review Letters* 59: 381.

Bang, C., J. L. Sabo, and S. H. Faeth. 2010. "Reduced Wind Speed Improves Plant Growth in a Desert City." *PLOS ONE* 5: e11061. doi:10.1371/journal.pone.0011061.

Barabási, A. L. 2005. "Taming Complexity." *Nature Physics* 1, no. 2: 68–70.

Barabási, A. L., and Z. N. Oltvai. 2004. "Network Biology: Understanding the Cell's Functional Organization." *Nature Reviews Genetics* 5: 101–13.

Barbour, J. 2014. "The Mystery of Time's Arrow." *Nautilus: The Future of Time* 9, no. 4. http://nautil.us/issue/9/time/the-mystery-of-times-arrow.

Barnosky A. D., E. A. Hadly, J. Bascompte, E. L. Berlow, J. H. Brown, M. Fortelius, M. W. Getz, J. Harte, A. Hastings, P. A. Marquet, N. D. Martinez, A. Mooers, P. G. Roopnarine Vermeij, J. W. Williams, R. Gillespi, G. Kitzes, C. Marshall, N. Matzke, D. P. Mindell, E. Revilla, A. B. Smith, 2012. "Approaching a state shift in Earth's biosphere." *Nature* 486: 52-58.

Barrington-Leigh, C., and A. Millard-Ball. 2015. "A Century of Sprawl in the

United States." *Proceedings of the National Academy of Sciences* 112, no. 27: 8244–49. doi:10.1073/pnas.1504033112.

Bascompte, J., P. Jordano, and J. M. Olesen. 2006. "Asymmetric Coevolutionary Networks Facilitate Biodiversity Maintenance." *Science* 312: 431–33. doi:10.1126/science.1123412.

Batty, M. 2005. *Cities and Complexity: Understanding Cities through Cellular Automata, Agent-Based Models, and Fractals.* Cambridge, MA: MIT Press.

Batty, M., and Y. Xie. 1999. "Self-Organized Criticality and Urban Development." *Discrete Dynamics in Nature and Society* 3, nos. 2–3: 109–24. doi: 10.1155/S1026022699000151.

Beatley, T. 2010. *Biophilic Cities: Integrating Nature into Urban Design and Planning.* Washington, DC: Island Press.

Bengston, N. N., G. H. Kubik, and P. C. Bishop. 2012. "Strengthening Environmental Foresight: Potential Contributions of Futures Research." *Ecology and Society* 17, no. 2: 10. doi:10.5751/ES-04794-170210.

Berglund, N., and B. Gentz. 2002. "Metastability in Simple Climate Models: Pathwise Analysis of Slowly Driven Langevin Equations." *Stochastic Dynamics* 2: 327–56.

Bernstein, Jeremy. 1989. "Besso." *New Yorker* (February 27): 86–87.

Bettencourt, L. M. A. 2013. "The Origins of Scaling in Cities." *Science* 340, no. 6139: 1438–41. doi:10.1126/science.1235823.

Bettencourt, L. M. A., J. Lobo, D. Helbing, C. Kuhnert, and G. B. West. 2007. "Growth, Innovation, Scaling, and the Pace of Life in Cities." *Proceedings of the National Academy of Sciences* 104, no. 17: 7301–06. doi:10.1073/pnas .0610172104.

Bettencourt, L. M. A., and G. West. 2010. "A Unified Theory of Urban Living." *Nature* 467: 912–13.

Biesbroek, G. R., C. J. A. M. Termeer, J. E. M. Klostermann, and P. Kabat. 2013. "On the Nature of Barriers to Climate Change Adaptation." *Regional Environmental Change* 13, no. 5: 1119–29.

Biggs, R., T. Blenckner, C. Folke, L. J. Gordon, A. Norström, M. Nyström, and G. D. Peterson. 2012. "Regime Shifts." In *Sourcebook in Theoretical Ecology*, edited by A. Hastings and L. Gross, 609–17. Berkeley: University of California Press.

Biondi, F. 2014. "Paleoecology Grand Challenge." *Frontiers in Ecology and Evolution* 2, no. 50. http://dx.doi.org/10.3389/fevo.2014.00050.

Booth, D. B. 1990. "Stream Channel Incision Following Drainage-Basin Urbanization." *Water Resources Bulletin* 26: 407–17.

Booth, D. B., and C. J. Fischenich. 2015. "A Channel Evolution Model to Guide Sustainable Urban Stream Restoration." *Area* 47, no. 4: 408–21. doi:10.1111 /area.12180.

Booth, D. B., and C. J. Jackson. 1997. "Urbanization of Aquatic Systems: Degradation Thresholds, Stormwater Detention, and the Limits of Mitigation." *Water Resources Bulletin* 33: 1077–90.

Brock, W. A., and S. R. Carpenter. 2006. "Variance as a Leading Indicator of Regime Shift in Ecosystem Services." *Ecology and Society* 11, no. 2: 9.

Brock, W. A., S. R. Carpenter, and M. Scheffer. 2008. "Regime Shifts, Environmental Signals, Uncertainty, and Policy Choice." In *Complexity Theory for a Sustainable Future*, edited by J. Norberg and G. Cumming, 180–206. New York: Columbia University Press.

Brock, W. A., S. N. Durlauf, and G. Rondina. 2013. "Design Limits and Dynamic Policy Analysis." *Journal of Economic Dynamics and Control* 37, no. 12: 2710 –28. doi:10.1016/j.jedc.2013.07.008.

Brooks, H. 1986. "The Typology of Surprises in Technology, Institutions and Development." In *Sustainable Development of the Biosphere*, edited by W. C. Clark and R. E. Munn, 235–48. Cambridge: Cambridge University Press.

Brown, J. H. 1995. *Macroecology.* Chicago: University of Chicago Press.

Brown, J. H., V. K. Gupta, B.-L. Li, B. T. Milne, C. Restrepo, and G. B. West. 2002. "The Fractal Nature of Nature: Power Laws, Ecological Complexity and Biodiversity." *Philosophical Transactions of the Royal Society of London, Series B* (357): 619–26.

Burnside, W. R., J. H. Brown, O. Burger, M. J. Hamilton, M. Moses, and L. Bettencourt. 2012. "Human Macroecology: Linking Pattern and Process in Big-Picture Human Ecology." *Biological Reviews* 87: 194–208. doi:10.1111/j.1469 -185X.2011.00192.x.

Cadenasso, M. L., and S. T. A. Pickett. 2013. "Three Tides: The Development and State of the Art of Urban Ecological Science." In *Resilience in Ecology and Urban Design*, edited by S. T. A. Pickett, M. L. Cadenasso, and B. P. McGrath, 29–46. Dordrecht: Springer.

Calvino, I. 1974 [1972]. *Invisible Cities.* New York: Harcourt Brace.

Canadell, J., R. Dickenson, K. Hibbard, M. Raupach, and O. Young. 2003. *Global Carbon Project: Science Framework and Implementation.* Global Carbon Project Report no. 1. Canberra: Earth System Science Partnership (IGBP, IHDP, WCRP, DIVERSITAS).

Cardinale, B. J., H. Hillebrand, and D. F. Charles. 2006. "Geographic Patterns of Diversity in Streams are Predicted by a Multivariate Model of Disturbance and Productivity." *Journal of Ecology* 94, no. 3: 609–18.

Cardinale, B. J., K. Nelson, and M. A. Palmer. 2000. "Linking Species Diversity to the Functioning of Ecosystems: On the Importance of Environmental Context." *Oikos* 91: 175–83.

Cardinale, B. J., and M. A. Palmer. 2002. "Disturbance Moderates Biodiversity-Ecosystem Function Relationships: Experimental Evidence from Caddisflies in Stream Mesocosms." *Ecology* 83, no. 7: 1915–27.

Carlson, S. M., T. P. Quinn, and A. P. Hendry. 2011. "Eco-Evolutionary Dynamics in Pacific Salmon." *Heredity* 106: 438–47.

Carpenter, S. R. 2002. "Ecological Futures: Building an Ecology of the Long Now." *Ecology* 83: 2069–83.

Carpenter, S. R., E. G. Booth, S. Gillon, C. J. Kucharik, S. Loheide, A. S. Mase,

M. Motew, J. Qiu, A. R. Rissman, J. Seifert, E. Soylu, M. Turner, and C. B. Wardropper. 2015a. "Plausible Futures of a Social-Ecological System: Yahara Watershed, Wisconsin, USA." *Ecology and Society* 20, no. 2: 10. http://dx.doi.org/10.5751/es-07433-200210.

Carpenter, S. R., and W. A. Brock. 2004. "Spatial Complexity, Resilience, and Policy Diversity: Fishing on Lake-Rich Landscapes." *Ecology and Society* 9, no. 1: 8.

————. 2006. "Rising Variance: A Leading Indicator of Ecological Transition." *Ecology Letters* 9: 311–18.

Carpenter, S. R., W. A. Brock, C. Folke, E. H. van Nes, and M. Scheffer. 2015b. "Allowing Variance May Enlarge the Safe Operating Space for Exploited Ecosystems." *Proceedings of the National Academy of Sciences* 112, no. 46: 14384–89. doi:10.1073/pnas.1511804112.

Carpenter, S. R., and L. H. Gunderson. 2001. "Coping with Collapse: Ecological and Social Dynamics in Ecosystem Management." *Bioscience* 51, no. 6: 451–57.

Carpenter, S. R., B. H. Walker, J. M. Anderies, and N. Abel. 2001. "From metaphor to measurement: resilience of what to what?" *Ecosystems* 4: 765–81.

Carlson, J., and J. Doyle. 2000. "Highly Optimized Tolerance: Robustness and Design in Complex Systems." *Physical Review Letters* 84, no. 11: 2529–32.

Cash, D. W., W. C. Clark, F. Alcock, N. Dickinson, N. Eckley, D. Guston, J. Jäger, and R. Mitchell. 2003. "Knowledge Systems for Sustainable Development." *Proceedings of the National Academy of Sciences of the USA* 100, no. 14: 8086–91.

Changizi, M. A., M. A. McDannald, and D. Widders. 2002. "Scaling of Differentiation in Networks: Nervous Systems, Organisms, Ant Colonies, Ecosystems, Businesses, Universities, Cities, Electronic Circuits, and Legos." *Journal of Theoretical Biology* 218: 215–37.

Chapin III, F. S., C. Folke, and G. P. Kofinas. 2009. "A Framework for Understanding Change." In *Principles of Ecosystem Stewardship*, edited by F. S. Chapin III, G. P. Kofinas, and C. Folke, 3–28. New York: Springer.

Chapin III, F. S., M. E. Power, S. T. A. Pickett, A. Freitag, J. A. Reynolds, R. B. Jackson, D. M. Lodge, et al. 2011. "Earth Stewardship: Science for Action to Sustain the Human-Earth System." *Ecosphere* 2, no. 8: art89. doi:10.1890/ES11-00166.1.

Chapin III, F. S., and A. M. Starfield. 1997. "Time Lags and Novel Ecosystems in Response to Transient Climatic Change in Arctic Alaska." *Climatic Change* 35: 449–61.

Chelleri, L., J. J. Waters, M. Olazabal, and G. Minucci. 2015. "Resilience Trade-Offs: Addressing Multiple Scales and Temporal Aspects of Urban Resilience." *Environment and Urbanization*. 0956247814550780. doi:10.1177/0956247814550780.

Cheptou, P. O., O. Carrue, S. Rouifed, and A. Cantarel. 2008. "Rapid Evolution of Seed Dispersal in an Urban Environment in the Weed *Crepis sancta.*" *Proceedings of the National Academy of Sciences of the USA* 105: 3796–99.

Churkina, G., D. G. Brown, and G. Keoleian. 2010. "Carbon Stored in Human Settlements: The Conterminous United States." *Global Change Biology* 16, no. 1: 135–43.

Clark, J. S., S. R. Carpenter, M. Barber, S. Collins, A. Dobson, J. Foley, D. Lodge, et al. 2001. "Ecological Forecasting: An Emerging Imperative." *Science* 293: 657–60.

Clark, W. C. 1986. "Sustainable Development of the Biosphere: Themes for a Research Program." In *Sustainable Development of the Biosphere*, edited by C. Clark and R. E. Munn, 4–48. Cambridge: Cambridge University Press.

Climate Central, Inc. 2012. *Climate Central: Global Weirdness: Severe Storms, Deadly Heat Waves, Relentless Drought, Rising Seas, and the Weather of the Future.* New York: Pantheon Books.

Collins, J., A. Kinzig, N. Grimm, W. Fagan, D. Hope, J. Wu, and E. Borer. 2000. "A New Urban Ecology." *American Scientist* 88, no. 5: 416–25. doi:10.1511/2000.5.416.

Collins, S., and G. Bell. 2004. "Phenotypic Consequences of 1,000 Generations of Selection at Elevated CO_2 in a Green Alga." *Nature* 431: 566–69.

Collins, W. J., S. Sitch, and O. Boucher. 2010. "How Vegetation Impacts Affect Climate Metrics for Ozone Precursors." *Journal of Geophysical Research* 115, no. D23308. doi:10.1029/2010JD014187.

Connell, J. H. 1978. "Diversity in Tropical Rain Forests and Coral Reefs." *Science* 199: 1302–10.

Connor, E. F., and E. D. McCoy. 1979. "The Statistics and Biology of the Species-Area Relationship." *American Naturalist* 113: 791–833.

Cook, C. N., S. Inayatullah, M. A. Burgman, W. J. Sutherland, and B. A. Wintle. 2014. "Strategic Foresight: How Planning for the Unpredictable Can Improve Environmental Decision-Making." *Trends in Ecology and Evolution* 29, no. 9: 531–41.

Cook, E., S. J. Hall, and K. Larson. 2012. "Residential Landscapes in an Urban Socio-Ecological Context: A Synthesis of Multi-Scalar Interactions between People and Their Home Environment." *Urban Ecosystems* 15: 1–52.

Cook, R. E. 2000. "Do Landscapes Learn? Ecology's New Paradigm and Design in Landscape Architecture." In *Environmentalism in Landscape Architecture*, edited by M. C. Conan, 115–32. Washington, DC: Dumbarton Oaks Trustees for Harvard Library.

Cooke, S. J., C. D. Suski, K. G. Ostrand, D. H. Wahl, and D. P. Philipp. 2007. "Physiological and Behavioral Consequences of Long-Term Artificial Selection for Vulnerability to Recreational Angling in a Teleost Fish." *Physiological and Biochemical Zoology* 80, no. 5: 480–90.

Coreau, A., G. Pinay, J. D. Thompson, P. O. Cheptou, and L. Mermet. 2009.

"The Rise of Research on Futures in Ecology: Rebalancing Scenarios and Predictions." *Ecology Letters* 12, no. 12: 1277–86.

Coreau, A., S. Treyer, P. O. Cheptou, J. D. Thompson, and L. Mermet. 2010. "Exploring the Difficulties of Studying Futures in Ecology: What Do Ecological Scientists Think?" *Oikos* 119, no. 8: 1364–76.

Costanza, R., B. S. Low, E. Ostrom, and J. Wilson, eds. 2001. *Institutions, Ecosystems, and Sustainability*. Boca Raton, FL: Lewis Publishers.

Crutzen, P. J. 2002. "Geology of Mankind: The Anthropocene." *Nature* 415: 23.

Crutzen, P. J., and E. F. Stoermer. 2000. "The 'Anthropocene.'" *IGBP Newsletter* 41: 17–18.

Cumming, G. S., D. H. M. Cumming, and C. L. Redman. 2006. "Scale Mismatches in Social-Ecological Systems: Causes, Consequences, and Solutions." *Ecology and Society* 11, no. 1: 14.

Daily, G. C., S. Alexander, P. R. Ehrlich, L. Goulder, J. Lubchenco, J. Matson, P. A. Mooney, et al. 1997. "Environmental Services: Benefits Supplied to Human Societies by Natural Ecosystems." *Issues in Ecology* 2, no. 16, www.esa.org/esa/wp-content/uploads/2013/03/issue2.pdf.

Dearlove, D. 2002. "Peter Schwartz, Thinking the Unthinkable: An Interview with Peter Schwartz, Scenario Planning Futurist." *Business*, September 22–23.

DeFries, R. S., E. C. Ellis, F. S. Chapin III, P. A. Matson, B. L. Turner II, A. Agrawal, P. J. Crutzen, et al. 2012. "Planetary Opportunities: A Social Contract for Global Change Science to Contribute to a Sustainable Future." *BioScience* 62, no. 6: 603–06.

Deheyn, D. D., and L. R. Shaffer. 2007. "Saving Venice: Engineering and Ecology in the Venice Lagoon." *Technology in Society* 29: 205–13.

Dent, C. L., G. S. Cumming, and S. R. Carpenter. 2002. "Multiple States in River and Lake Ecosystems." *Philosophical Transactions of the Royal Society of London Series B: Biological Sciences* 357: 635–45.

Desrochers, A. 2010. "Morphological Response of Songbirds to 100 Years of Landscape Change in North America." *Ecology* 91: 1577–82.

Dewey, J. 1939. *Logic: The Theory of Inquiry*. New York: Henry Holt.

Diamond, J. 2005. *Collapse: How Societies Choose to Fail or Succeed*. New York: Viking.

Dirzo, R., H. S. Young, M. Galetti, G. Ceballos, N. J. B. Isaac, and B. Collen. 2014. "Defaunation in the Anthropocene." *Science* 345, no. 6195: 401–06.

Donihue, C. M., and M. R. Lambert. 2015. "Adaptive Evolution in Urban Ecosystems." *Ambio: A Journal of the Human Environment* 44, no. 3: 194–203. doi:10.1007/s13280-014-0547-2.

Dover, G. A., and R. B. Flavell. 1984. "Molecular Coevolution: DNA Divergence and the Maintenance of Function." *Cell* 38: 622–23.

Dowd, M., and R. Meyer. 2003. "A Bayesian Approach to the Ecosystem Inverse Problem." *Ecological Modelling* 168: 39–55.

Dowling, J. L., D. A. Luther, and P. P. Marra. 2011. "Comparative Effects of

Urban Development and Anthropogenic Noise on Bird Songs." *Behavioral Ecology* 23, no. 1: 201–09.

Dunne, T., and L. B. Leopold. 1978. *Water in Environmental Planning*. San Francisco: W. H. Freeman.

Durham, W. H. 1991. *Coevolution: Genes, Culture and Human Diversity*. Stanford, CA: Stanford University Press.

Edmondson, W. T. 1991. *Uses of Ecology: Lake Washington and Beyond*. Seattle: University of Washington Press.

Ehrenberg, R. 2015. "Urban Microbes Come Out of the Shadows." *Nature* 522: 399–400. doi:10.1038/522399a.

Ehrman, J. R., and B. L. Stinson. 1999. "Joint Fact-Finding and the Use of Technical Experts." In *The Consensus Building Handbook*, edited by L. Susskind, S. McKearnan and J. Thomas-Larmer, 375–99. Thousand Oaks, CA: Sage.

Elton, C. S. 1927. *Animal Ecology*. Chicago: University of Chicago Press.

Embarq. 2013. *Sustainable Transport Adoption Curves*. Washington, DC: World Resources Institute.

Endler, J. A. 1986. *Natural Selection in the Wild*. Princeton, NJ: Princeton University Press.

Ernstson, H., S. E. van der Leeuw, C. L. Redman, D. J. Meffert, G. Davis, C. Alfsen, and T. Elmqvist. 2010. "Urban Transitions: On Urban Resilience and Human-Dominated Ecosystems." *Ambio: A Journal of the Human Environment* 39, no. 8: 531–45. doi:10.1007/s13280-010-0081-9.

Faeth, S. H., P. S. Warren, E. Shochat, and W. A. Marussich. 2005. "Trophic Dynamics in Urban Communities." *BioScience* 55: 399. doi:10.1641/0006 -3568(2005)055[0399:TDIUC]2.0.CO;2.

Falkowski, P., R. J. Scholes, E. Boyle, J. Canadell, D. Canfield, J. Elser, N. Gruber, et al. 2000. "The Global Carbon Cycle: A Test of Our Knowledge of Earth as a System." *Science* 290, no. 5490: 291–96.

Fauchald, P., and T. Tveraa. 2006. "Hierarchical Patch Dynamics and Animal Movement Pattern." *Oecologia* 149, no. 3: 383–95.

Felson, A. J., M. A. Bradford, and E. Oldfield. 2013. "Involving Ecologists in Shaping Large-Scale Green Infrastructure Projects." *BioScience* 63, no. 11: 882–90.

Firestein, S. 2012. *Ignorance: How It Drives Science*. New York: Oxford University Press.

Florida, R. 2013. *The Mega-Regions of North America*. Toronto: University of Toronto, Martin Prosperity Institute. http://martinprosperity.org/media /Mega%20Regions_Insight_14-03-05.pdf.

Florida, R., C. Mellander, and T. Gulden. 2012. "Global Metropolis: Assessing Economic Activity in Urban Centers Based on Nighttime Satellite Images." *Professional Geographer* 64, no. 2: 178–87.

Folke, C. 2006. "Resilience: The Emergence of a Perspective for Social-Ecological Systems Analyses." *Global Environmental Change* 16, no. 3: 253–67.

Folke, C., S. Carpenter, T. Elmqvist, L. Gunderson, C. S. Holling, B. Walker, J. Bengtsson, et al. 2002. *Resilience and Sustainable Development: Building Adaptive Capacity in a World of Transformation*. Report for the Swedish Environmental Advisory Council 2002:1. Stockholm: Ministry of the Environment.

Folke, C., S. Carpenter, B. Walker, M. Scheffer, T. Elmqvist, L. Gunderson, and C. S. Holling. 2004. "Regime Shifts, Resilience, and Biodiversity in Ecosystem Management." *Annual Review of Ecology, Evolution, and Systematics* 35: 557–81. doi:10.1146/annurev.ecolsys.35.021103.105711.

Forman, R. T. T. 2008. *Urban Regions: Ecology and Planning Beyond the City*. Cambridge and New York: Cambridge University Press.

———. 2014. *Urban Ecology: Science of Cities*. Cambridge: Cambridge University Press.

Frank, A. 2013. "The City as Infestation" (radio program). NPR. Cosmos and Culture, 13.7. October 9, 2012. www.npr.org/blogs/13.7/2012/10/09/162506048/the-city-as-infestation.

Frank, A., M. Alberti and A. Kleidon. 2015. "Earth's Transition into a Hybridized State: The Anthropocene in an Evolutionary Astrobiological Context." Unpublished ms.

Franssen, N. R., L. K. Stewart, and J. F. Schaefer. 2013. "Morphological Divergence and Flow-Induced Phenotypic Plasticity in a Native Fish from Anthropogenically Altered Stream Habitats." *Ecology and Evolution* 3: 4648–57.

Frumkin, H. 2013. "The Evidence of Nature and the Nature of Evidence." *American Journal of Preventive Medicine* 44, no. 2: 196–97.

Fussmann, G. F., M. Loreau, and P. A. Abrams. 2007. "Eco-Evolutionary Dynamics of Communities and Ecosystems." *Functional Ecology* 21: 465–77.

Gaertner, M., R. Biggs, M. Te Beest, C. Hui, J. Molofsky, and D. M. Richardson. 2014. "Invasive Plants as Drivers of Regime Shifts: Identifying High-Priority Invaders That Alter Feedback Relationships." *Diversity and Distributions* 20, no. 7: 733–44.

Geddes, P. 1915. *City in Evolution*. London: Williams and Norgate.

Georgescu, M., P. E. Morefield, B. G. Bierwagen, and C. P. Weaver. 2014. "Urban Adaptation can Roll Back Warming of Emerging Megapolitan Regions." *Proceedings of the National Academy of Sciences* 111, no. 8: 2909–14. doi:10.1073/pnas.1322280111.

Gilbert, O. L. 1989. *The Ecology of Urban Habitats*. New York: Chapman and Hall.

Gisiger, T. 2001. "Scale Invariance in Biology: Coincidence or Footprint of a Universal Mechanism?" *Biological Reviews of the Cambridge Philosophical Society* 76, no. 2: 161–209.

Glaeser, E. 2012. *Triumph of the City: How Our Greatest Invention Makes Us Richer, Smarter, Greener, Healthier, and Happier*. New York: Penguin Books.

———. 2013. "A World of Cities: The Causes and Consequences of Urbanization in Poorer Countries." National Bureau of Economic Research, Working Paper no. 19745. Cambridge, MA: NBER.

Gleason, H. A.1926. "The Individualistic Concept of the Plant Association." *Bulletin of the Torrey Botanical Club* 53: 7–27.

Godet, M. 1986. "Introduction to *la prospective*: Seven Key Ideas and One Scenario Method." *Futures* 18, no. 2: 134–57.

Goldblatt, C., T. M. Lenton, and A. J. Watson. 2006. "Bistability of Atmospheric Oxygen and the Great Oxidation." *Nature* 443, no. 7112: 683–86. doi:10.1038/nature05169.

Goldenfeld, N., and L. P. Kadanoff. 1999. "Simple Lessons from Complexity." *Science* 284, 5411: 87–89.

Gomez-Ramirez, J. 2013. "Don't Blame the Economists. It Is an Inverse Problem!" *European Journal of Futures Research* 1, no. 13. doi:10.1007/s40309 -013-0013-6.

Gomez-Ramirez, J., and R. Sanz. 2013. "On the Limitations of Standard Statistical Modeling in Biological Systems: A Full Bayesian Approach for Biology." *Progress in Biophysics and Molecular Biology*. doi:10.1016/j.pbiomolbio.2013 .03.008.

Gottmann, J. 1961. *Megalopolis: The Urbanized Northeastern Seaboard of the United States*. New York: Twentieth Century Fund.

Gould, S. J. 1980. "The Promise of Paleobiology as a Nomothetic, Evolutionary Discipline." *Paleobiology* 6: 96–118.

Goyal, S. 2015. "Networks in Economics: A Perspective on the Literature." In *Oxford Handbook on Economics of Networks*, edited by Y. Bramoulle, A. Galeotti, and B. Rogers Cambridge: INET Institute.

Graham, M. H., and P. K. Dayton. 2002. "On the Evolution of Ecological Ideas: Paradigms and Scientific Progress." *Ecology* 83, no. 6: 1481–89.

Grimm, N. B., F. S. Chapin III, B. Bierwagen, P. Gonzalez, P. M. Groffman, Y. Luo, F. Melton, et al. 2013. "The Impacts of Climate Change on Ecosystem Structure and Function." *Frontiers in Ecology and the Environment* 11, no. 9: 474–82.

Grimm, N. B., S. H. Faeth, N. E. Golubiewski, C. L. Redman, J. Wu, X. M. Bai, and J. M. Briggs. 2008a. "Global Change and the Ecology of Cities." *Science* 319, no. 5864: 756–60.

Grimm, N. B., D. Foster, P. Groffman, J. M. Grove, C. S. Hopkinson, K. J. Nadelhoffer, D. E. Pataki, and D. P. Peters. 2008b. "The Changing Landscape: Ecosystem Responses to Urbanization and Pollution across Climatic and Societal Gradients." *Frontiers in Ecology and the Environment* 6: 264–72.

Grimm, N. B., J. M. Grove, S. T. A. Pickett, and C. L. Redman. 2000. "Integrated Approaches to Long-Term Studies of Urban Ecological Systems." *BioScience* 50, no. 7: 571–84.

Grimmond, C. S. B. 2006. "Progress in Measuring and Observing the Urban Atmosphere." *Theoretical and Applied Climatology* 84, nos. 1–3: 3–22.

Groffman, P. M., K. Butterbach-Bahl, R. W. Fulweiler, A. J. Gold, J. L. Morse,

E. K. Stander, C. Tague, C. Tonitto, and P. Vidon. 2009. "Challenges to Incorporating Spatially and Temporally Explicit Phenomena (Hotspots and Hot Moments) in Denitrification Models." *Biogeochemistry* 93, no. 1: 49–77.

Groffman, P. M., J. Cavender-Bares, N. D. Bettez, J. Morgan Grove, S. J. Hall, J. B. Heffernan, S. E. Hobbie, et al. 2014. "Ecological Homogenization of Urban USA." *Frontiers in Ecology and the Environment* 12, no. 1: 74–81.

Groffman, P. M., and M. K. Crawford. 2003. "Denitrification Potential in Urban Riparian Zones." *Journal of Environmental Quality* 32, no. 3: 1144–49.

Grose, M. J. 2014. "Thinking Backwards can Inform Concerns about 'Incomplete' Data." *Trends in Ecology and Evolution* 29, no. 10: 546–47. doi:10.1016/j.tree.2014.07.007.

Gunderson, L. H., and C. S. Holling, eds. 2002. *Panarchy: Understanding Transformations in Human and Natural Systems.* Washington, DC: Island Press.

Gurney, K. R., D. L. Mendoza, Y. Zhou, M. L. Fischer, C. Miller, S. Geethakumar, and S. de la Rue du Can. 2009. "High Resolution Fossil Fuel Combustion CO_2 Emission Fluxes for the United States." *Environmental Science and Technology* 43, no. 14: 5535–41.

Haas, T. C., M. J. Blum, and D. C. Heins. 2010. "Morphological Responses of a Stream Fish to Water Impoundment." *Biology Letters* 6: 803–06.

Hairston, N. G., C. L. Holtmeier, W. Lampert, L. J. Weider, D. M. Post, J. M. Fischer, C. E. Caceres, J. A. Fox, and U. Gaedke. 2001. "Natural Selection for Grazer Resistance to Toxic Cyanobacteria: Evolution of Phenotypic Plasticity?" *Evolution* 55: 2203–14.

Hall, P. 2001. "Global City-Regions in the Twenty-First Century." In *Global City-Regions: Trends, Theory, Policy*, edited by A. J. Scott, 59–77. Oxford: Oxford University Press.

Hamilton, M. J., B.-T. Milne, R. S. Walker, O. Burger, and J. H. Brown. 2007. "The Complex Structure of Hunter-Gatherer Social Networks." *Proceedings of the Royal Society of London B: Biological Sciences* 274, no. 1622: 2195–2203.

Hanna, K., and D. S. Slocombe, eds. 2007. *Integrated Resource and Environmental Management: Concepts and Practice.* Oxford and Toronto: Oxford University Press.

Hansen, A. J., R. L. Knight, J. M. Marzluff, S. Powell, K. Brown, P. H. Gude, and K. Jones. 2005. "Effects of Exurban Development on Biodiversity: Patterns, Mechanisms, and Research Needs." *Ecological Applications* 15, no. 6: 1893–1905.

Harris, S. E., J. Munshi-South, C. Obergfell, and R. O'Neill. 2013. "Signatures of Rapid Evolution in Urban and Rural Transcriptomes of White-Footed Mice (*Peromyscus leucopus*) in the New York Metropolitan Area." *PLOS ONE* 8, no. 8: e74938. doi:10.1371/journal.pone.0074938.

Harrison, J., and M. Hoyler. 2014. *Megaregions: Globalization's New Urban Form?* London: Edward Elgar.

Hartvigsen, G., A. Kinzig, and G. Peterson. 1998. "Use and Analysis of Complex Adaptive Systems in Ecosystem Science: Overview of Special Section." *Ecosystems* 1: 427–30.

Harvey, D. 1989. *The Condition of Postmodernity: An Enquiry into the Origins of Cultural Change*. New York: Blackwell.

Haskell, J. P., M. E. Ritchie, and H. Olff. 2002. "Fractal Geometry Predicts Varying Body Size Scaling Relationships for Mammal and Bird Home Ranges." *Nature* 418: 527–30.

Hayes, J. R. 1989. *The Complete Problem Solver*. 2nd ed. Hillsdale, NJ: Erlbaum.

He, F., and P. Legendre. 1996. "On Species-Area Relations." *American Naturalist* 148: 719–37.

Heffernan, J. B., R. A. Sponseller, and S. G. Fisher. 2008. "Consequences of a Biogeomorphic Regime Shift for the Hyporheic Zone of a Sonoran Desert Stream." *Freshwater Biology* 53: 1954–68.

Hegger, D., M. Lamers, A. Van Zeijl-Rozema, and C. Dieperink. 2012 "Conceptualising Joint Knowledge Production in Regional Climate Change Adaptation Projects: Success Conditions and Levers for Action." *Environmental Science and Policy* 18: 52–65.

Helbing, D. 2013. "Globally Networked Risks and how to Respond." *Nature* 497: 51–59.

Hendry, A. P., T. J. Farrugia, and M. T. Kinnison. 2008. "Human Influences on Rates of Phenotypic Change in Wild Animal Populations." *Molecular Ecology* 17: 20–29.

Hendry, A. P., and M. T. Kinnison. 1999. "The Pace of Modern Life: Measuring Rates of Contemporary Microevolution." *Evolution* 53: 1637–53.

Hines, A. 2006. "Strategic Foresight: The State of the Art." *Futurist* 40, no. 5: 18–21.

Hirsch, P. D., and B. G. Norton. 2012. "Thinking like a Planet." In *Ethical Adaptation to Climate Change: Human Virtues of the Future*, edited by A. Thompson and J. Bendik-Keymer, 317. Cambridge, MA: MIT Press.

Hirsch Hadorn, G., S. Biber-Klemm, and W. Grossenbacher-Mansuy. 2008. "The Emergence of Transdisciplinarity as a Form of Research." In *Handbook of Transdisciplinary Research*, edited by G. Hirsch Hadorn, H. Hoffmann-Riem, S. Biber-Klemm, W. Grossenbacher-Mansuy, D. Joye, C. Pohl, U. Wiesmann, and E. Zemp, 19–39. Dordrecht: Springer.

Hixon, M. A., S. W. Pacala, and S. A. Sandin. 2002. "Population Regulation: Historical Context and Contemporary Challenges of Open vs. Closed Systems." *Ecology* 83, no. 6: 1490–508.

Hobbs, R. J., S. Arico, J. Aronson, J. S. Brown, P. Bridgewater, V. A. Cramer, P. R. Epstein, et al. 2006. "Novel Ecosystems: Theoretical and Management Aspects of the New Ecological World Order." *Global Ecology and Biogeography* 15: 1–7.

Hobbs, R. J., E. Higgs, and J. A. Harris. 2009. "Novel Ecosystems: Implications for Conservation and Restoration." *Trends in Ecology and Evolution* 24, no. 11: 599–605. doi:10.1016/j.tree.2009.05.012.

Holling, C. S. 1973. "Resilience and Stability of Ecological Systems." *Annual Review of Ecology and Systematics* 4, no. 1: 1–23.

————. 1986. "The Resilience of Terrestrial Ecosystems; Local Surprise and Global Change." In *Sustainable Development of the Biosphere*, edited by W. C. Clark and R. E. Munn. 292–317. Cambridge: Cambridge University Press.

————. 1992. "Cross-Scale Morphology, Geometry, and Dynamics of Ecosystems." *Ecological Monographs* 62: 447–502.

————. 1995. "What Barriers? What Bridges?" In *Barriers and Bridges to the Renewal of Ecosystems and Institutions*, edited by L. Gunderson, C. S. Holling, and S. Light, 3–34. New York: Columbia University Press.

————. 1996. "Surprise for Science, Resilience for Ecosystems, and Incentives for People." *Ecological Applications* 6: 733–35.

Holling, C. S., and L. H. Gunderson. 2002. "Resilience and Adaptive Cycles." In *Panarchy: Understanding Transformations in Human and Natural Systems*, edited by L. H. Gunderson and C. S. Holling, 25–62. Washington, DC: Island Press.

Holling, C. S., L. H. Gunderson, and G. D. Peterson. 2002. "Sustainability and Panarchies." In *Panarchy: Understanding Transformations in Human and Natural Systems*, edited by L. H. Gunderson and C. S. Holling, 63–102. Washington, DC: Island Press.

Hopkins, K. G., N. B. Morse, D. J. Bain, N. D. Bettez, N. B. Grimm, J. L. Morse, and M. M. Palta. 2015. "Type and Timing of Stream Flow Changes in Urbanizing Watersheds in the Eastern US." *Elementa: Science of the Anthropocene* 3, no. 1: 000056.

Howard, E. 1902. *Garden Cities of To-Morrow*. London: Swan Sonnenschein and Co.

Hughes, A. R., J. E. Byrnes, D. L. Kimbro, and J. J. Stachowicz. 2007. "Reciprocal Relationships and Potential Feedbacks between Biodiversity and Disturbance." *Ecology Letters* 10, no. 9: 849–64. doi:10.1111/j.1461-0248.2007.01075.

Hughes, T. P., A. H. Baird, E. Dinsdale, N. Moltschaniwskyj, M. S. Pratchett, J. E. Tanner, and B. L. Willis. 2012. "Assembly Rules of Reef Corals Are Flexible along a Steep Climatic Gradient." *Current Biology* 22, no. 8: 736–41.

Hughes, T. P., S. Carpenter, J. Rockström, M. Scheffer, and B. Walker. 2013. "Multiscale Regime Shifts and Planetary Boundaries." *Trends in Ecology and Evolution* 28, 389–95.

Hume, D. 1777 [1748]. *An Enquiry Concerning Human Understanding*. London: A. Millar.

Hutchinson, G. E. 1957. "Concluding Remarks." *Cold Spring Harbor Symposium on Quantitative Biology* 22: 415–27.

————. 1965. *The Ecological Theater and the Evolutionary Play*. New Haven: Yale University Press.

Hutyra, L. R., R. Duren, K. R. Gurney, N. Grimm, E. A. Kort, E. Larson, and G. Shrestha. 2014. "Urbanization and the Carbon Cycle: Current Capabilities and Research Outlook from the Natural Sciences Perspective." *Earth's Future* 2: 473–95. doi:10.1002/2014EF000255.

Hutyra, L. R., B. Yoon, and M. Alberti. 2011. "Terrestrial Carbon Stocks across a Gradient of Urbanization: A Study of the Seattle, WA Region." *Global Change Biology* 17, no. 2: 783–87.

Iglesias, M., and A. M. Stuart. 2014. "Inverse Problems and Uncertainty Quantification." *SIAM NEWS*, 2–3.

Inayatullah, S. 2005. *Questioning the Future: Methods and Tools for Organizational and Societal Transformation*. Taipei: Tamkang University Press.

Innes, J. E., and D. E. Booher. 1999. "Metropolitan Development as a Complex System: A New Approach to Sustainability." *Economic Development Quarterly* 13: 141–56.

IPCC (Intergovernmental Panel on Climate Change). 2012. "Summary for Policymakers." In *Managing the Risks of Extreme Events and Disasters to Advance Climate Change Adaptation*, edited by C. B. Field, V. Barros, T. F. Stocker, D. Qin, D. J. Dokken, K. L. Ebi, M. D. Mastrandrea, et al., 1–19. Special Report of Working Groups I and II of the Intergovernmental Panel on Climate Change. Cambridge and New York: Cambridge University Press.

———. 2014a. *Climate Change 2014: Impacts, Adaptation, and Vulnerability. Part A: Global and Sectoral Aspects*, edited by C. B. Field, V. R. Barros, D. J. Dokken, K. J. Mach, M. D. Mastrandrea, T. E. Bilir, M. Chatterjee, et al. Contribution of Working Group II to the Fifth Assessment Report of the Intergovernmental Panel on Climate Change. Cambridge and New York: Cambridge University Press.

———. 2014b. *Climate Change 2014: Synthesis Report*, edited by R. K. Pachauri and L A. Meyer. Contribution of Working Groups I, II, and III to the Fifth Assessment Report of the Intergovernmental Panel on Climate Change. Geneva: IPCC.

Ives, A. R., and S. R. Carpenter. 2007. "Stability and Diversity of Ecosystems." *Science* 317, no. 5834: 58–62.

Jacobs, J. 1961. *The Death and Life of Great American Cities*. New York: Random House.

Jacobson, R. B., S. R. Femmer, and R. A. McKenney. 2001. "Land-Use Changes and the Physical Habitat of Streams: A Review with Emphasis on Studies within the US Geological Survey Federal-State Cooperative Program." *USGS Circular* 1175.

Jacquemyn, H., L. De Meester, E. Jongejans, and O. Honnay. 2012. "Evolutionary Changes in Plant Reproductive Traits Following Habitat Fragmentation and Their Consequences for Population Fitness." *Journal of Ecology* 100, no. 1: 76–87.

Janssen, M., O. Bodin, and J. Anderies. 2006. "Toward a Network Perspective of the Study of Resilience in Social-Ecological Systems." *Ecology and Society* 11, no. 1: 32.

Jasanoff, S., and M. L. Martello, eds. 2004. *Earthly Politics: Local and Global in Environmental Governance*. Cambridge, MA: MIT Press.

Johnson, B. R., and S. Kwan Lam. 2010. "Self-Organization, Natural Selection, and Evolution: Cellular Hardware and Genetic Software." *BioScience* 60: 879–85.

Jones, R. N., and B. L. Preston. 2011. "Adaptation and Risk Management." *Wiley Interdisciplinary Reviews: Climate Change* 2, no. 2: 296–308. doi:10.1002 /wcc.97.

Julien, P. Y., and C. W. Vensel. 2005. *Review of Sediment Issues on the Mississippi River*. Fort Collins: Department of Civil and Environmental Engineering, Colorado State University.

Kabanikhin, S. I. 2008. "Definitions and Examples of Inverse and Ill-Posed Problems." *Journal of Inverse and Ill-Posed Problems* 16, no. 4: 317–57. doi: 10.1515 / JIIP.2008.069.

Kambhu, J., S. Weidman, and N. Krishnan. 2007. *New Directions for Understanding Systemic Risk*. National Academies Press, Washington DC.

Kauffman, S. A. 1993. *The Origins of Order: Self-Organization and Selection in Evolution*. Oxford: Oxford University Press.

———. 1995. *At Home in the Universe: The Search for Laws of Self-Organization and Complexity*. Oxford and New York: Oxford University Press.

Kauffman, S. A., and S. Johnsen. 1991. "Co-Evolution to the Edge of Chaos: Coupled Fitness Landscapes, Poised States, and Co-Evolutionary Avalanches." *Journal of Theoretical Biology* 149: 467–505.

Kawecki, T. J., and D. Ebert. 2004. "Conceptual Issues in Local Adaptation." *Ecology Letters* 7, no. 12: 1225–41.

Kay, J. 2010. *Obliquity*. London: Profile Books.

Kaye, J. P., P. M. Groffman, N. B. Grimm, L. A. Baker, and R. V. Pouyat. 2006. "A Distinct Urban Biogeochemistry?" *Trends in Ecology and Evolution* 21: 192–99.

Kerkhoff, A. J., and B. J. Enquist. 2007. "Implications of Scaling Approaches for Understanding Resilience and Reorganization in Ecosystems." *BioScience* 57: 489–99.

Kettlewell, H. B. N. 1958. "A Survey of the Frequencies of *Biston Betularia* (L) (Lep.) and Its Melanic Forms in Great Britain." *Heredity* 12: 51–72.

Kille, P., J. Andre, C. Anderson, H. N. Ang, M. W. Bruford, J. G. Bundy, R. Donnelly, et al. 2013. "DNA Sequence Variation and Methylation in an Arsenic Tolerant Earthworm Population." *Soil Biology and Biochemistry* 57: 524–32.

King, D. A., and S. M. Thomas. 2007. "Taking Science out of the Box: Foresight Recast." *Science* 316, no. 5832: 1701–02.

King, G. 1997. *A Solution to the Ecological Inference Problem: Reconstructing Individual Behavior from Aggregate Data*. Princeton, NJ: Princeton University Press.

Kleiber, M. 1932. "Body Size and Metabolism." *Hilgardia* 6: 315–51.

Klein, R. D. 1979. "Urbanization and Stream Quality Impairment." *Journal of the American Water Resources* Association 15, no. 4: 948–63.

Kolbert, E. 2015. *The Sixth Extinction: An Unnatural History.* New York: Picador.

Kowarik, I. 2011. "Novel Urban Ecosystems, Biodiversity, and Conservation." *Environmental Pollution* 159: 1974–83.

Krebs, C. J. 1988. *The Message of Ecology.* New York: Harper & Row.

Krugman, P. 1996. *The Self-Organizing Economy.* Cambridge, MA, and Oxford: Blackwell.

Kühn, I., R. Brandl, and S. Klotz. 2004. "The Flora of German Cities Is Naturally Species Rich." *Evolutionary Ecology Research* 6: 749–64.

Laland, K. N. 2014. "On Evolutionary Causes and Evolutionary Processes." *Behavioral Processes* doi:10.1016/j.beproc.2014.05.008.

Laland, K. N., K. Sterelny, J. Odling-Smee, W. Hoppitt, and T. Uller. 2011. "Cause and Effect in Biology Revisited: Is Mayr's Proximate-Ultimate Dichotomy Still Useful?" *Science* 334: 1512–16.

Landry, C. R., D. L. Hartl, and J. M. Ranz. 2007. "Genome Clashes in Hybrids: Insights from Gene Expression." *Heredity* 99: 483–93.

Lawton, J. H. 1994. "What Do Species Do in Ecosystems?" *Oikos* 71: 367–74.

Le Corbusier. 1935. *La Ville radieuse* [The Radiant City]. Boulogne: Editions de l'Architecture d'aujourd'hui.

Lenton, T. M. 1998. "Gaia and Natural Selection." *Nature* 394, no. 6692: 439–47.

Lenton, T. M., and H. T. Williams. 2013. "On the Origin of Planetary-Scale Tipping Points." *Trends in Ecology and Evolution* 28, no. 7: 380–82.

Leopold, A. 1949. *A Sand County Almanac: And Sketches Here and There.* New York: Oxford University Press.

Levin, S. A. 1992. "The Problem of Pattern and Scale in Ecology." *Ecology* 73, no. 6: 1943–67.

———. 1998. "Ecosystems and the Biosphere as Complex Adaptive Systems." *Ecosystems* 1, no. 5: 431–36.

———. 1999. *Fragile Dominion: Complexity and the Commons.* Cambridge: Perseus.

Liao, K. 2012. "A Theory on Urban Resilience to Floods: A Basis for Alternative Planning Practices." *Ecology and Society* 17, no. 4: 48.

Likens, G. E. 1992. *The Ecosystem Approach: Its Use and Abuse.* Excellence in Ecology, Book 3. Oldendorf-Luhe, Germany: Ecology Institute.

Liu, J., T. Dietz, S. R. Carpenter, C. Folke, M. Alberti, C. L. Redman, S. H. Schneider, et al. 2007b. "Coupled Human and Natural Systems." *Ambio: A Journal of the Human Environment* 36, no. 8: 639–49.

Liu, J., V. Hull, M. Batistella, R. Defries, T. Dietz, F. Fu, T. W. Hertel, et al. 2013. "Framing Sustainability in a Telecoupled World." *Ecology and Society* 18, no. 2: 26. doi:10.5751/ES-05873-180226.

Liu, J., H. Mooney, V. Hull, S. J. Davis, J. Gaskell, T. Hertel, J. Lubchenco, et al. 2015. "Systems Integration for Global Sustainability." *Science* 347, no. 6225: 1258832. doi:10.1126/science.1258832.

Liu, J. G., T. Dietz, S. R. Carpenter, M. Alberti, C. Folke, E. Moran, A. N. Pell, et al. 2007a. "Complexity of Coupled Human and Natural Systems." *Science* 317: 1513–16.

Loreau, M. 2010. "Linking Biodiversity and Ecosystems: Towards a Unifying Ecological Theory." *Philosophical Transactions of the Royal Society of London B: Biological Sciences* 365, no. 1537: 49–60.

Loreau, M., and C. de Mazancourt. 2013. "Biodiversity and Ecosystem Stability: A Synthesis of Underlying Mechanisms." *Ecological Letters* 16: 106–15.

Lovelock, J. 1995. *The Ages of Gaia: A Biography of Our Living Earth.* Oxford and New York: Norton.

Lowe, E. C., S. M. Wilder, and D. F. Hochuli. 2014. "Urbanisation at Multiple Scales Is Associated with Larger Size and Higher Fecundity of an Orb-Weaving Spider." *PLOS ONE* 9, no. 8: e105480.

Lowe, J., T. Reeder, K. Horsburgh, and V. Bell. 2008. *Using the New TE2100 Science Scenarios.* Rotherham: UK Environment Agency.

Lynch, K. 1972. *What Time Is This Place?* Cambridge, MA: MIT Press.

———. 1981. *A Theory of Good City Form.* Cambridge, MA: MIT Press.

MacArthur, R. H. 1969. "Patterns of Communities in the Tropics." *Biological Journal of the Linnaean Society of London* 1: 19–30.

MacArthur, R. H., and E. O. Wilson. 1963. "An Equilibrium Theory of Insular Zoogeography." *Evolution* 17: 373–87.

Machlis, G. E., J. E. Force, and W. R. Burch. 1997. "The Human Ecosystem Part I: The Human Ecosystem as an Organizing Concept in Ecosystem Management." *Society and Natural Resources: An International Journal* 10, no. 4: 347–67.

Marani, M. 2013. "Venice: Lessons Learned on Resilience and the Natural Environment." Presentation at "Symposium on Urban Resilience in an Era of Climate Change," New York, October 17–18, 2013.

Marleau, J. N., F. Guichard, and M. Loreau. 2014. "Meta-Ecosystem Dynamics and Functioning on Finite Spatial Networks." *Proceedings of the Royal Society of London B* 218, no. 1777 (February). doi:10.1098/rspb.2013.2094.

Marquet, P. A., A. P. Allen, J. H. Brown, J. A. Dunne, B. J. Enquist, J. F. Gillooly, P. A. Gowaty, et al. 2014. "On Theory in Ecology." *BioScience* 64: 701–10. doi:10.1093/biosci/biu098.

Marsh, G. P. 1864. *Man and Nature, Or Physical Geography as Modified by Human Action.* London: Sampson Low, Son and Marston.

Martin, H., and N. Goldenfeld. 2006. "On the Origin and Robustness of Power-Law Species-Area Relationships in Ecology." *Proceedings of the National Academy of Sciences of the USA* 103: 310–15.

Marx, B., T. Stoker, and T. Suri. 2013. "The Economics of Slums in the Developing World." *Journal of Economic Perspectives.* 187–210. doi:10.1257/jep .27.4.187.

Marzluff, J. M. 2005. "Island Biogeography for an Urbanizing World: How Extinction and Colonization May Determine Biological Diversity in Human-Dominated Landscapes." *Urban Ecosystems* 8: 155–75.

———. 2008. "Island Biogeography for an Urbanizing World: How Extinction and Colonization May Determine Biological Diversity in Human-

Dominated Landscapes." In *Urban Ecology*, edited by J. M. Marzluff,
E. Shulenberger, W. Endlicher, M. Alberti, G. Bradley, C. Ryan, U. Simon,
and C. ZumBrunnen, 355–71. New York: Springer.

―――. 2012. "Urban Evolutionary Ecology." *Studies in Avian Biology* 45:
287–308.

Marzluff, J. M., J. C. Withey, K. A. Whittaker, M. D. Oleyar, T. M. Unfried,
S. Rullman, and J. Delap. 2007. "Consequences of Habitat Utilization
by Nest Predators and Breeding Songbirds across Multiple Scales in an
Urbanizing Landscape." *Condor* 109: 516. doi:10.1650/8349.1.

Mascaro, J., J. A. Harris, L. Lach, A. Thompson, M. P. Perring, D. M. Richard-
son, and E. Ellis. 2013. "Origins of the Novel Ecosystems Concept." In *Novel
Ecosystems: Intervening in the New Ecological World Order*, edited by V. R. J.
Hobbs, E. S. Higgs, and C. M. Hall, 45–57. Chichester: Wiley and Sons.

Matthews, B., A. Narwani, S. Hausch, E. Nonaka, H. Peter, M. Yamamichi,
K. E. Sullam, et al. 2011. "Toward an Integration of Evolutionary Biology
and Ecosystem Science." *Ecology Letters* 14: 690–701.

May, R. M., S. A. Levin, and G. Sugihara. 2008. "Complex Systems: Ecology
for Bankers." *Nature* 451: 893–95. doi:10.1038/451893a.

McDonnell, M. J., and S. Pickett, eds. 1993. *Humans as Components of Ecosystems: The Ecology of Subtle Human Effects and Populated Areas*. New York:
Springer.

McDonnell, M. J., S. T. A. Pickett, P. Groffman, P. Bohlen, R. V. Pouyat, W. C.
Zipperer, R. W. Parmelee, M. M. Carreiro, and K. Medley. 1997. "Ecosystem
Processes along an Urban-to-Rural Gradient." *Urban Ecosystems* 1: 21–36.

McGrath, B., and D. G. Shane. 2012. "Introduction: Metropolis, Megalopolis
and the Metacity." In *The Sage Handbook of Architectural Theory*, edited by
C. G. Crysler, S. Cairns, and H. Heynen, 641–57. London: Sage Publications.
doi:http://dx.doi.org/10.4135/9781446201756.n38.

McHale, M. R., S. T. A. Pickett, O. Barbosa, N. N. Bunn, M. L. Cadenasso, D. L.
Childers, M. Gartin, et al. 2015. "The New Global Urban Realm: Complex,
Connected, Diffuse, and Diverse Social-Ecological Systems." *Sustainability*
7, no. 5: 5211–40.

McIntyre, P. B., A. S. Flecker, M. J. Vanni, J. M. Hood, B. W. Taylor, and S. A.
Thomas. 2008. "Fish Distributions and Nutrient Cycling in Streams: Can
Fish Create Biogeochemical Hotspots?" *Ecology* 89: 2335–46.

McPhearson, T., S. T. A. Pickett, N. B. Grimm, J. Niemelä, M. Alberti,
T. Elmqvist, C. Weber, D. Haase, J. Breuste, and S. Qureshi. 2016. "Advancing Urban Ecology Towards a Science of Cities." *BioScience* doi: 10.1093/
biosci/biw002.

McKinney, M. L. 2006. "Urbanization as a Major Cause of Biotic Homogenization." *Biological Conservation* 127, no. 3: 247–60.

Millennium Ecosystem Assessment (MEA). 2005. *Ecosystems and Human Well-
Being: Synthesis*. Washington, DC: Island Press.

Miller, A. W., A. C. Reynolds, C. Sobrino, and G. F. Riedel. 2009. "Shellfish Face

Uncertain Future in High CO$_2$ World: Influence of Acidification on Oyster Larvae Calcification and Growth in Estuaries." *PLOS ONE* 4, no. 5: e5661. doi:10.1371/journal.pone.0005661.

Miller, F., H. Osbahr, E. Boyd, F. Thomalla, S. Bharwani, G. Ziervogel, B. Walker, et al. 2010. "Resilience and Vulnerability: Complementary or Conflicting Concepts." *Ecology and Society* 15, no. 3: 11.

Milne, B. T. 1998. "Motivation and Benefits of Complex Systems Approaches in Ecology." *Ecosystems* 1: 449–56.

Milne, B. T., M. G. Turner, J. A. Wiens, and A. R. Johnson. 1992. "Interactions between the Fractal Geometry of Landscapes and Allometric Herbivory." *Theoretical Population Biology* 41: 337–53.

Milton, S. J. 2003. "Emerging Ecosystems: A Washing Stone for Ecologists, Economists and Sociologists?" *South African Journal of Science* 99: 404–06.

Miner, B. E., L. De Meester, M. E. Pfrender, W. Lampert, and N. G. Hairston. 2012. "Linking Genes to Communities and Ecosystems: *Daphnia* as an Eco-genomic Model." *Proceedings of the Royal Society of London B* 279: 1873–82.

Miranda, A. C., H. Schielzeth, T. Sonntag, and J. Partecke. 2013. "Urbanization and Its Effects on Personality Traits: A Result of Microevolution or Pheno-typic Plasticity?" *Global Change Biology* 19: 2634–44.

Mittelbach, G. G., C. F. Steiner, S. M. Scheiner, K. L. Gross, H. L. Reynolds, R. B. Waide, M. R. Willig, S. I. Dodson, and L. Gough. 2001. "What Is the Observed Relationship between Species Richness and Productivity?" *Ecology* 82: 2381–96.

Monastersky, R. 2015. "Anthropocene: The Human Age." *Nature* 519, no. 7542: 144–47. www.nature.com/news/anthropocene-the-human-age-1.17085.

Mori, A. S., T. Furukawa, and T. Sasaki. 2013. "Response Diversity Determines the Resilience of Ecosystems to Environmental Change." *Biological Reviews* 88: 349–64.

Moser, S. C., and J. A. Ekstrom. 2010. "A Framework to Diagnose Barriers to Climate Change Adaptation." *Proceedings of the National Academy of Science* 107: 22026–31.

Mulder, C. P. H., D. D. Uliassi, and D. F. Doak. 2001. "Physical Stress and Diversity-Productivity Relationships: The Role of Positive Interactions." *Proceedings of the National Academy of Science* 98, no. 12: 6704–08.

Nash, K. L., C. R. Allen, D. G. Angeler, C. Barichievy, T. Eason, A. S. Garmes-tani, N. A. J. Graham, et al. 2014. "Discontinuities, Cross-Scale Patterns, and the Organization of Ecosystems." *Ecology* 95, no. 3: 654–67. http://dx.doi.org/10.1890/13-1315.1.

Nash, K. L., N. A. J. Graham, S. K. Wilson, and D. R. Bellwood. 2013. "Cross-Scale Habitat Structure Drives Fish Body Size Distributions on Coral Reefs." *Ecosystems* 16, no. 3: 478–90. doi:10.1007/s10021-012-9625-0.

Newman, M. E. 2006. "Modularity and Community Structure in Networks." *Proceedings of the National Academy of Sciences* 103, no. 23: 8577–82. doi: 10.1073/pnas.0601602103.

New York City. 2011. *PlaNYC April 2011 Update: A Greener, Greater New York.*
New York: Mayor's Office, http://nytelecom.vo.llnwd.net/o15/agencies
/planyc2030/pdf/planyc_2011_planyc_full_report.pdf.

————. 2013. *PlaNYC: A Stronger, More Resilient New York.* New York: NYEDC,
www.nycedc.com/resource/stronger-more-resilient-new-york.

New York City Panel on Climate Change (NPCC). 2010. *Climate Change Adaptation in New York City: Building a Risk Management Response,* vol. 1196. New
York: New York Academy of Sciences.

Nicolis, G., and I. Prigogine. 1977. *Self-Organization in Nonequilibrium Systems.*
New York: Wiley.

————. 1989. *Exploring Complexity: An Introduction.* New York: W. H. Freeman.

Norberg, J., and G. S. Cumming. 2008. *Complexity Theory for a Sustainable Future.*
New York: Columbia University Press.

Nowak, D. J., and D. E. Crane. 2002. "Carbon Storage and Sequestration by
Urban Trees in the USA." *Environmental Pollution* 116, no. 3: 381–89.

O'Connor, T., and H. Y. Wong. 2015. "Emergent Properties." In *The Stanford
Encyclopedia of Philosophy,* edited by E. N. Zalta. Stanford, CA: Metaphysics
Research Lab, Center for the Study of Language and Information. http://
plato.stanford.edu/archives/sum2015/entries/properties-emergent/.

Odling-Smee, F. J. 2003. *Niche Construction: The Neglected Process in Evolution.*
Princeton, NJ: Princeton University Press.

Odling-Smee, F. J., K. N. Laland, and M. W. Feldman. 2003. *Niche Construction: The Neglected Process in Evolution.* Princeton, NJ: Princeton University
Press.

Odum, E. P. 1971. *Fundamentals of Ecology.* 3rd ed. Philadelphia: Saunders.

————. 1975. *Ecology: The Link between the Natural and Social Sciences.* 2nd ed.
New York: Holt, Rinehart, and Winston.

Ohtsuki, H., C. Hauert, E. Lieberman, and M. A. Nowak. 2006. "A Simple Rule
for the Evolution of Cooperation on Graphs and Social Networks." *Nature*
441, no. 7092: 502–05.

Oke, T. R. 1982. "The Energetic Basis of the Urban Heat Island." *Quarterly Journal of the Royal Meteorological Society* 108, no. 455: 1–24.

Olden, J. D. 2008. "Biotic Homogenization." In *Encyclopedia of Life Sciences.*
Chichester, UK: John Wiley and Sons Ltd. doi:10.1002/9780470015902.
a0020471.

Olden, J. D., and N. L. Poff. 2004. "Ecological Processes Driving Biotic Homogenization: Testing a Mechanistic Model Using Fish Faunas." *Ecology* 85,
no. 7: 1867–75.

————. 2001. "Is It Time to Bury the Ecosystem Concept? (With Full Military
Honors, of Course!)" *Ecology* 82, no. 12: 3275–84.

O'Neill, R. V., D. L. DeAngelis, J. B. Waide, and T. F. H. Allen. 1986. *A Hierarchical Concept of Ecosystems.* Monographs in Population Biology 23. Princeton,
NJ: Princeton University Press.

O'Neill, R. V., R. H. Gardner, L. W. Barnthouse, G. W. Suter, S. G. Hildebrand,

and C. W. Gehrs. 1982. "Ecosystem Risk Analysis: A New Methodology." *Environmental Toxicology and Chemistry* 1, no. 2: 167–77.

Ortman, S. G., A. H. F. Cabaniss, J. O. Sturm, and L. M. A. Bettencourt. 2015. "Settlement Scaling and Increasing Returns in an Ancient Society." *Science Advances* 1, no. 1: e1400066. doi:10.1126/sciadv.1400066.

Ostrom, E. 1996. "Crossing the Great Divide: Coproduction, Synergy, and Development." *World Development* 24, no. 6: 1073–87.

———. 2009. "A General Framework for Analyzing Sustainability of Social-Ecological Systems." *Science* 325: 419–22.

Oteros-Rozas, E., B. Martín-López, T. Daw, E. L. Bohensky, J. Butler, R. Hill, J. Martin-Ortega, A. Quinlan, F. Ravera, I. Ruiz-Mallén, M. Thyresson, J. Mistry, I. Palomo, G. D. Peterson, T. Plieninger, K. A. Waylen, D. Beach, I. C. Bohnet, M. Hamann, J. Hanspach, K. Hubacek, S. Lavorel, and S. Vilardy. 2015. "Participatory Scenario Planning in Place-based Social-Ecological Research: Insights and Experiences from 23 Case Studies." *Ecology and Society* 20, no. 4: 32. http://dx.doi.org/10.5751/ES-07985-200432.

Overeem, I., and J. P. M. Syvitski. 2009. *Dynamics and Vulnerability of Delta Systems*. LOICZ Reports and Studies no. 35. Geesthacht, Germany: GKSS Research Center.

Ozawa, C. P. 1991. *Recasting Science: Consensual Procedures*. Boulder, CO: Westview Press.

Ozawa, C. P., and L. Susskind. 1985. "Mediating Science-Intensive Disputes." *Journal of Policy Analysis and Management* 5, no. 1: 23–29.

Pachauri, R. K., M. R. Allen, V. R. Barros, J. Broome, W. Cramer, R. Christ, J. A. Church, et al. 2014. *Climate Change 2014: Synthesis Report: Contribution of Working Groups I, II and III to the Fifth Assessment Report of the Intergovernmental Panel on Climate Change*, edited by R. Pachauri and L. Meyer. Geneva, Switzerland: IPCC.

Padisák, J. 1992. "Seasonal Succession of Phytoplankton in a Large Shallow Lake (Balaton, Hungary): A Dynamic Approach to Ecological Memory, Its Possible Role and Mechanisms." *Journal of Ecology* 80: 217–30.

Paine, R. T. 1984. "Ecological Determinism in the Competition for Space." *Ecology* 65: 1339–48.

Palkovacs, E. P., and A. P. Hendry. 2010. "Eco-Evolutionary Dynamics: Intertwining Ecological and Evolutionary Processes in Contemporary Time." *F1000 Biology Reports* 2, no. 1. doi:10.3410/B2-1.

Palkovacs, E. P., M. T. Kinnison, C. Correa, C. M. Dalton, and A. P. Hendry. 2012. "Fates Beyond Traits: Ecological Consequences of Human-Induced Trait Change." *Evolutionary Applications* 5, no. 2: 183–91.

Palumbi, S. R. 2001. "Humans as the World's Greatest Evolutionary Force." *Science* 293: 1786–90.

Pan, W., G. Ghoshal, C. Krumme, M. Cebrian, and A. Pentland. 2013. "Urban Characteristics Attributable to Density-Driven Tie Formation." *Nature Communications* 4: 1–35.

Parandehgheibi, M., and E. Modiano. 2013. "Robustness of Interdependent Networks: The Case of Communication Networks and the Power Grid." In *Global Communications Conference* (GLOBECOM), IEEE conference proceedings, December 9–13, 2013, 2164–69. http://hdl.handle.net/1721.1/96988.

Parris, K. M., and D. L. Hazell. 2005. "Biotic Effects of Climate Change in Urban Environments: The Case of the Grey-Headed Flying Fox (*Pteropus poliocephalus*) in Melbourne, Australia." *Biological Conservation* 124: 267–76.

Partecke, J. 2014. "Mechanisms of Phenotypic Responses following Colonization of Urban Areas: From Plastic to Genetic Adaptation." In *Avian Urban Ecology: Behavioural and Physiological Adaptations*, edited by D. Gil and H. Brumm. Oxford: Oxford University Press.

Pataki, D., M. Alberti, M. L. Cadenasso, A. J. Felson, M. McDonnell, S. Pincetl, R. V. Pouyat, H. Setälä, and T. Whitlow. 2015. "Are the Emperor's New Clothes Green? Why Urban Greening Needs More Science." Unpublished ms.

Pataki, D. E., R. J. Alig, A. S. Fung, N. E. Golubiewski, C. A. Kennedy, E. G. McPherson, D. J. Nowak, R. V. Pouyat, and P. Romero Lankao. 2006. "Urban Ecosystems and the North American Carbon Cycle." *Global Change Biology* 12, no. 11: 2092–102.

Pataki, D. E., T. Xu, Y. S. Q. Luo, and J. R. Ehleringer. 2007. "Inferring Biogenic and Anthropogenic Carbon Dioxide Sources across an Urban to Rural Gradient." *Oecologia* 152, no. 2: 307–22.

Paul, M. J., and J. L. Meyer. 2001. "Streams in the Urban Landscape." *Annual Review of Ecological Systematics* 32: 333–65.

Pearl, J. 2009. *Causality: Models, Reasoning, and Inference.* 2nd ed. Cambridge: Cambridge University Press.

Pelletier, F., D. Garant, and A. P. Hendry. 2009. "Eco-Evolutionary Dynamics." *Philosophical Transactions of the Royal Society of London B: Biological Sciences* 364: 1483–89.

Peters, R. H. 1983. *The Ecological Implications of Body Size.* New York: Cambridge University Press.

Peterson, G. D. 2002. "Contagious Disturbance, Ecological Memory, and the Emergence of Landscape Patterns." *Ecosystems* 5, no. 4: 329–38.

Peterson, G. D., C. R. Allen, and C. S. Holling. 1998. "Ecological Resilience, Biodiversity, and Scale." *Ecosystems* 1: 6–18.

Peterson, G. D., G. S. Cumming, and S. R. Carpenter. 2003. "Scenario Planning: A Tool for Conservation in an Uncertain World." *Conservation Biology* 17: 358–66. doi:10.1046/j.1523-1739.2003.01491.x.

Petraitis, P. S., R. E. Latham, and R. A. Niesenbaum. 1989. "The Maintenance of Species Diversity by Disturbance." *Quarterly Review of Biology*: 393–418.

Petri, G., P. Expert, H. J. Jensen, and J. W. Polak. 2013. "Entangled Communities and Spatial Synchronization Lead to Criticality in Urban Traffic." *Scientific Reports* 3, no. 1798. doi:10.1038/srep01798.

Philipp, D. P., S. J. Cooke, J. E. Claussen, J. B. Koppelman, C. D. Suski, and

D. P. Burkett. 2009. "Selection for Vulnerability to Angling in Largemouth Bass." *Transactions of the American Fisheries Society* 138: 189–99.

Pickett, S. T. A., W. R. Burch, S. E. Dalton, and T. W. Foresman. 1997a. "Integrated Urban Ecosystem Research." *Urban Ecosystems* 1: 183–84. doi:10 .1023/A:1018579628818.

Pickett, S. T. A., W. R. Burch, S. E. Dalton, T. W. Foresman, and J. M. Grove. 1997b. "A Conceptual Framework for the Study of Human Ecosystems in Urban Areas." *Urban Ecosystems* 1: 185–99.

Pickett, S. T. A., and M. L. Cadenasso. 2002. "The Ecosystem as a Multidimensional Concept: Meaning, Model, and Metaphor." *Ecosystems* 5: 1–10.

Pickett, S. T. A., M. L. Cadenasso, J. M. Grove, C. G. Boone, P. M. Groffman, E. Irwin, S. S. Kaushal, et al. 2011. "Urban Ecological Systems: Scientific Foundations and a Decade of Progress." *Journal of Environmental Management* 92: 331–62. doi:10.1016/j.jenvman.2010.08.22.

Pickett, S. T. A., M. L. Cadenasso, J. M. Grove, P. M. Groffman, L. E. Band, C. G. Boone, W. R. Burch, et al. 2008. "Beyond Urban Legends: An Emerging Framework of Urban Ecology, as Illustrated by the Baltimore Ecosystem Study." *BioScience* 58, no. 2: 139.

Pickett, S. T. A., M. L. Cadenasso, J. M. Grove, C. H. Nilon, R. V. Pouyat, W. C. Zipperer, and R. Costanza. 2001. "Urban Ecological Systems: Linking Terrestrial, Ecological, Physical, and Socioeconomic Components of Metropolitan Areas." *Annual Review of Ecology and Systematics* 32: 127–57.

Pickett, S. T. A., M. L. Cadenasso, and C. G. Jones. 2000. "Generation of Heterogeneity by Organisms: Creation, Maintenance, and Transformation." In *Ecological Consequences of Habitat Heterogeneity*, edited by M. Hutchings, L. John, and A. Stewart, 33–52. New York: Blackwell.

Pickett, S. T. A., M. L. Cadenasso, and B. P. McGrath. 2013. *Resilience in Urban Ecology and Urban Design: Linking Theory and Practice for Sustainable Cities.* New York: Springer.

Pickett, S. T. A., J. Kolasa, and C. G. Jones. 2010. *Ecological Understanding: The Nature of Theory and the Theory of Nature.* New York: Academic Press.

Pickett, S. T. A., V. T. Parker, and P. Fiedler. 1992. "The New Paradigm in Ecology: Implications for Conservation Biology above the Species Level." In *Conservation Biology: The Theory and Practice of Nature Conservation, Preservation, and Management,* edited by P. L. Fiedler and S. K. Jain, 65–88. New York: Chapman and Hall.

Pickett, S. T. A., and K. Rogers. 1997. "Patch Dynamics: The Transformation of Landscape Structure and Function." In *Wildlife and Landscape Ecology,* edited by J. Bissonnette, 101–27. New York: Springer.

Pickett, S. T. A, and P. S. White, eds. 2013. *The Ecology of Natural Disturbance and Patch Dynamics.* New York: Elsevier.

Pickett, S. T. A., J. Wu, and M. L. Cadenasso. 1999. "Patch Dynamics and the Ecology of Disturbed Ground: A Framework for Synthesis." In *Ecosystems of the World: Ecosystems of Disturbed Ground*, edited by L. R. Walker, 707–22. Amsterdam: Elsevier Science.

Pickett, S. T. A., and W. Zhou. 2015. "Global Urbanization as a Shifting Context for Applying Ecological Science Toward the Sustainable City." *Ecosystem Health and Sustainability* 1, no. 1: 5. http://dx.doi.org/10.1890/EHS14-0014.1.

Pierce, J. L., and A. L. Delbecq. 1977. "Organizational Structure, Individual Attitudes and Innovation." *Academy of Management Review* 2: 27–37.

Pimentel, D. 1961. "Animal Population Regulation by the Genetic Feedback Mechanism." *American Naturalist* 95: 65–79.

Polsky, C., J. M. Grove, C. Knudson, P. M. Groffman, N. Bettez, J. Cavender-Bares, S. J. Hall, et al. 2014. "Assessing the Homogenization of Urban Land Management with an Application to US Residential Lawn Care." *Proceedings of the National Academy of Sciences* 111, no. 12: 4432–37.

Portugali, J. 2000. "Spatial Cognitive Dissonance and Socio-Spatial Emergence in a Self-Organizing City." In *Self-Organization and the City*, 141–73. Berlin and Heidelberg: Springer.

Post, D. M., and E. P. Palkovacs. 2009. "Eco-Evolutionary Feedbacks in Community and Ecosystem Ecology: Interactions between the Ecological Theatre and the Evolutionary Play." *Philosophical Transactions of the Royal Society of London, Series B 364*: 1629–40.

Post, D. M., E. P. Palkovacs, E. G. Schielke, and S. I. Dodson. 2008. "Intraspecific Variation in a Predator Affects Community Structure and Cascading Trophic Interactions." *Ecology* 89: 2019–32.

Power, M. E., D. Tilman, J. A. Estes, B. A. Menge, W. J. Bond, L. S. Mills, D. Gretchen, et al. 1996. "Challenges in the Quest for Keystones." *BioScience* 46: 609–20.

Prigogine, I. 1978. "Time, Structure, and Fluctuations." *Science* 201, no. 4358: 777–85.

Pyšek, P. 1993. "Factors Affecting the Diversity of Flora and Vegetation in Central European Settlements." *Vegetatio* 106, no. 1: 89–100.

Ranger, N., T. Reeder, and J. Lowe. 2013. "Addressing 'Deep' Uncertainty over Long-Term Climate in Major Infrastructure Projects: Four Innovations of the Thames Estuary 2100 Project." *EURO Journal on Decision Processes* 1, nos. 3–4: 233–62.

Rebele, F. 1994. "Urban Ecology and Special Features of Urban Ecosystems." *Global Ecology and Biogeography Letters* 4: 173–87.

Reisinger, A., D. Wratt, and S. Allan. 2011. "The Role of Local Government in Adapting to Climate Change: Lessons from New Zealand." In *Climate Change Adaptation in Developed Nations*, edited by D. Ford and L. Berrang-Ford, 303–19. Dordrecht: Springer.

Resilience Alliance. 2015. "Resilience Alliance Glossary." www.resalliance.org/index.php/glossary. Accessed October 2015.

Resilience Alliance and Santa Fe Institute. 2004. "Thresholds and Alternate States in Ecological and Social-Ecological Systems." www.resalliance.org /index.php/thresholds_database. Accessed October 2015.

Restrepo, C., and N. Arango. 2008. "Discontinuities in the Geographical Range Size of North American Birds and Butterflies: A Biogeographical Test of the Textural Discontinuity Hypothesis." In C. Allen, G. Peterson, and C. S. Holling, eds., *Cross-Scale Structure and Discontinuities in Ecosystems and Other Complex Systems*, 101–35. New York: Columbia University Press.

Revi, A., D. E. Satterthwaite, F. Aragón-Durand, J. Corfee-Morlot, R. B. R. Kiunsi, M. Pelling, D. C. Roberts, and W. Solecki. 2014. "Urban Areas." In: *Climate Change 2014: Impacts, Adaptation, and Vulnerability. Part A: Global and Sectoral Aspects. Contribution of Working Group II to the Fifth Assessment Report of the Intergovernmental Panel on Climate Change*, edited by C. B. Field, V. R. Barros, D. J. Dokken, K. J. Mach, M. D. Mastrandrea, T. E. Bilir, M. Chatterjee, et al., 535–612. Cambridge and New York: Cambridge University Press.

Reznick, N. N., M. Mateos, and M. S. Springer. 2002. "Independent Origins and Rapid Evolution of the Placenta in the Fish Genus *Poeciliopsis*." *Science* 298, no. 5595: 1018–20.

Riba, M., M. Mayol, B. E. Giles, O. Ronce, E. Imbert, M. Van Der Velde, S. Chauvet, et al. 2009. "Darwin's Wind Hypothesis: Does It Work for Plant Dispersal in Fragmented Habitats?" *New Phytologist* 183: 667–77.

Richter-Boix, A., M. Quintela, M. Kierczak, M. Franch, and A. Laurila. 2013. "Fine-Grained Adaptive Divergence in an Amphibian: Genetic Basis of Phenotypic Divergence and the Role of Nonrandom Gene Flow in Restricting Effective Migration among Wetlands." *Molecular Ecology* 22, no. 5: 1322–40.

Ritchie, M. E. 1998. "Scale-Dependent Foraging and Patch Choice in Fractal Environments." *Evolutionary Ecology* 12, no. 3: 309–30.

Robinson, J. 2003. "Future Subjunctive: Backcasting as Social Learning." *Futures* 35, no. 8: 839–56.

Robles, C., and R. Desharnais. 2002. "History and Current Development of a Paradigm of Predation in Rocky Intertidal Communities." *Ecology* 83, no. 6: 1521–36.

Robson, T. M., V. A. Pancotto, A. L. Scopel, S. D. Flint, and M. M. Caldwell. 2005. "Solar UV-B Influences Microfaunal Community Composition in a Tierra del Fuego Peatland." *Soil Biology and Biochemistry* 37, no. 12: 2205–15.

Rocha, J. C., G. D. Peterson, and R. Biggs. 2015. "Regime Shifts in the Anthropocene: Drivers, Risks, and Resilience." *PLOS ONE* 10, no. 8: e0134639. doi:10.1371/journal.pone.0134639.

Rockström, J., W. Steffen, K. Noone, Å. Persson, F. S. Chapin III, E. F. Lambin, T. M. Lenton, et al. 2009. "A Safe Operating Space for Humanity." *Nature* 461: 472–75.

Roff, D. A. 1974. "Spatial Heterogeneity and the Persistence of Populations." *Oecologia* 15: 245–58.

Rolshausen, G., G. Segelbacher, K. A. Hobson, and H. Martin Schaefer. 2009. "Contemporary Evolution of Reproductive Isolation and Phenotypic Divergence in Sympatry along a Migratory Divide." *Current Biology* 19, no. 24: 2097–2101.

Rosenzweig, C. and W. Solecki. 2010. "Climate Change Adaptation in New York City: Building a Risk Management Response." *Annals of the New York Academy of Sciences* 1196: 5–6. doi:10.1111/j.1749-6632.2009.05397.x.

Rosenzweig, C., W. Solecki, A. DeGaetano, M. O'Grady, S. Hassol, and P. Grabhorn, eds. 2011. *Responding to Climate Change in New York State: The ClimAID Integrated Assessment for Effective Climate Change Adaptation.* Technical report. Albany: New York State Energy Research and Development Authority (NYSERDA), www.nyserda.ny.gov.

———. 2014. "Hurricane Sandy and Adaptation Pathways in New York: Lessons from a First-Responder City." *Global Environmental Change* 28: 395–408.

Ross, C. 2009. *Megaregions: Planning for Global Competitiveness.* Washington, DC: Island Press.

Sabatier, P. C. 2000. "Past and Future of Inverse Problems." *Journal of Mathematical Physics* 41: 4082–124.

Sassen, S. 2012. *Cities in a World Economy.* Washington, DC: Sage.

Sax, D. F., and S. D. Gaines. 2003. "Species Diversity: From Global Decreases to Local Increases." *Trends in Ecology and Evolution* 11: 561–66.

Scheffer, M. 2014. "The Forgotten Half of Scientific Thinking." *PNAS* 111, no. 17: 6119. doi:10.1073/pnas.1404649111.

Scheffer, M., and S. R. Carpenter. 2003. "Catastrophic Regime Shifts in Ecosystems: Linking Theory to Observation." *Trends in Ecology and Evolution* 18, no. 12: 648–56. doi:10.1016/j.tree.2003.09.002.

Scheffer, M., S. Carpenter, J. A. Foley, C. Folke, and B. Walker. 2001. "Catastrophic Shifts in Ecosystems." *Nature* 413, no. 6856: 591–96.

Scheffer, M., S. R. Carpenter, T. M. Lenton, J. Bascompte, W. Brock, V. Dakos, J. van de Koppel, et al. 2012. "Anticipating Critical Transitions." *Science* 338, no. 6105: 344–48. doi:10.1126/science.1225244.

Schindler, D. E., and R. Hilborn. 2015. "Prediction, Precaution, and Policy Under Global Change." *Science* 347, no. 6225: 953–54.

Schläpfer, M., L. M. A. Bettencourt, S. Grauwin, M. Raschke, R. Claxton, Z. Smoreda, G. B. West, and C. Ratti. 2014. "The Scaling of Human Interactions with City Size." *Journal of the Royal Society, Interface* 11, no. 98: 20130789. doi:10.1098/rsif.2013.0789.

Schoemaker, P. J. H. 1995. "Scenario Planning: A Tool for Strategic Thinking." *MIT Sloan Management Review* 36: 25–40.

Schoener, T. W. 2009. "Ecological Niche." In *The Princeton Guide to Ecology*, edited by S. A. Levin, S. R. Carpenter, H. Charles J. Godfray, A. P. Kinzig, M. Loreau, J. B. Losos, B. Walker, and D. S. Wilcove, 3. Princeton, NJ: Princeton University Press.

————. 2011. "The Newest Synthesis: Understanding the Interplay of Evolutionary and Ecological Dynamics." *Science* 331, no. 6016: 426–29.

Schwartz, P. 2005. *The Art of the Long View: Planning for the Future in an Uncertain World*. Chichester, PA: Doubleday.

Schwartz, P., and J. A. Ogilvy. 1998. "Plotting Your Scenarios." In *Learning from the Future*, edited by L. Fahey and R. M. Randall, 57–80. New York: John Wiley and Sons.

Seehausen, O. 2006. "Conservation: Losing Biodiversity by Reverse Speciation." *Current Biology* 16: R334–37.

Seto, K. C., B. Güneralp, and L. R. Hutyra. 2012. "Global Forecasts of Urban Expansion to 2030 and Direct Impacts on Biodiversity and Carbon Pools." *Proceedings of the National Academy of Sciences of the USA* 109, no. 40: 16083–88. doi:10.1073/pnas.1211658109.

Shackell, N. L., K. T. Frank, J. A. D. Fisher, B. Petrie, and W. C. Leggett. 2010. "Decline in Top Predator Body Size and Changing Climate Alter Trophic Structure in an Oceanic Ecosystem." *Proceedings of the Royal Society of London B* 277, no. 1686: 1353–60.

Shenoy, K., and P. H. Crowley. 2011. "Endocrine Disruption of Male Mating Signals: Ecological and Evolutionary Implications." *Functional Ecology* 25, no. 3: 433–48. doi:10.1111/j.1365-2435.2010.01787.x.

Sherlock, R. R., S. G. Sommer, Z. R. Khan, C. W. Wood, E. A. Guertal, J. R. Freney, C. O. Dawson, and K. C. Cameron. 2002. "Ammonia, Methane, and Nitrous Oxide Emission from Pig Slurry Applied to a Pasture in New Zealand." *Journal of Environmental Quality* 31, no. 5: 1491–1502.

Shochat, E., S. B. Lerman, J. M. Anderies, P. S. Warren, S. H. Faeth, and C. H. Nilon. 2010. "Invasion, Competition, and Biodiversity Loss in Urban Ecosystems." *BioScience* 60, no. 3: 199–208.

Shochat, E., P. S. Warren, S. H. Faeth, N. E. McIntyre, and D. Hope. 2006. "From Patterns to Emerging Processes in Mechanistic Urban Ecology." *Trends in Ecology and Evolution* 21, no. 4: 186–91. doi:10.1016/j.tree.2005.11.019.

Sim, A., S. N. Yaliraki, M. Barahona, and M. P. H. Stumpf. 2015. "Great Cities Look Small." *Journal of the Royal Society Interface* 12: 1–9.

Slabbekoorn, H. 2013. "Songs of the City: Noise-Dependent Spectral Plasticity in the Acoustic Phenotype of Urban Birds." *Animal Behaviour* 85: 1089–99.

Snieder, R., and J. Trampert. 1999. "Inverse Problems in Geophysics." In *Wavefield Inversion*, edited by A. Wirgin, 119–90. New York: Springer.

Solé, R. V., S. C. Manrubia, M. Benton, S. Kauffman, and P. Bak. 1999. "Criticality and Scaling in Evolutionary Ecology." *Trends in Ecology and Evolution* 14, no. 4: 156–60.

Solé, R. V., and J. M. Montoya. 2001. "Complexity and Fragility in Ecological Networks." *Proceedings of the Royal Society of London B* 268: 2039–45. doi:10.1098/rspb.2001.1767.

Sosa Torres, M. E., J. P. Saucedo-Vázquez, and P. M. H. Kroneck. 2015. "The

Rise of Dioxygen in the Atmosphere." In *Sustaining Life on Planet Earth: Metalloenzymes Mastering Dioxygen and Other Chewy Gases*, edited by P. M. H. Kroneck and M. E. Sosa Torres, 1–12. *Metal Ions in Life Sciences*, no. 15. Switzerland: Springer.

Sousa, C. D. 2001. "Contaminated Sites: The Canadian Situation in an International Context." *Journal of Environmental Management* 62, no. 2: 131–54.

Stafford Smith, M., L. Horrocks, A. Harvey, and C. Hamilton. 2011. "Rethinking Adaptation for a 4 °C World." *Philosophical Transactions of the Royal Society A: Mathematical, Physical and Engineering Sciences* 369: 196–216.

Stahl, A. E., and L. Feigenson. 2015. "Observing the Unexpected Enhances Infants' Learning and Exploration." *Science* 348, no. 6230: 91–94.

Stanley, E. H., S. M. Powers, and N. R. Lottig. 2010. "The Evolving Legacy of Disturbance in Stream Ecology: Concepts, Contributions, and Coming Challenges." *Journal of the North American Benthological Society* 29: 67–83.

Steffen, W., P. J. Crutzen, and J. R. McNeill. 2007. "The Anthropocene: Are Humans Now Overwhelming the Great Forces of Nature?" *Ambio: A Journal of the Human Environment* 36, no. 8: 614–21.

Steffen, W., K. Richardson, J. Rockström, S. E. Cornell, I. Fetzer, E. M. Bennett, R. Biggs, et al. 2015. "Planetary Boundaries: Guiding Human Development on a Changing Planet." *Science* 347, no. 6223, 1259855. doi:10.1126/science.1259855.

Stiling, P., D. C. Moon, M. D. Hunter, J. Colson, A. M. Rossi, G. J. Hymus, and B. G. Drake. 2003. "Elevated CO_2 Lowers Relative and Absolute Herbivore Density Across all Species of a Scrub-Oak Forest." *Oecologia* 134: 82–87.

Stockholm Resilience Centre. 2015. "Regime Shifts Database." www.regime shifts.org. Accessed October 2014.

Stockwell, C. A. 2003. "Contemporary Evolution Meets Conservation Biology." *Trends in Ecology and Evolution* 18: 94101.

Strauss, B., R. Ziemlinski, J. Weiss, and J. T. Overpeck. 2012. "Tidally-Adjusted Estimates of Topographic Vulnerability to Sea Level Rise and Flooding for the Contiguous United States." *Environmental Research Letters* 7 014033. doi:10.1088/1748-9326/7/1/014033.

Suding, K. N., and R. J. Hobbs. 2009. "Threshold Models in Restoration and Conservation: A Developing Framework." *Trends in Ecology and Evolution* 24: 271–79.

Sukopp, H., H. P. Blume, and W. Kunick. 1979. "The Soil, Flora and Vegetation of Berlin's Wastelands." In *Nature in Cities*, edited by I. C. Laurie, 115–32. Chichester, UK: John Wiley.

Susskind, L., and J. Cruikshank. 1987. *Breaking the Impasse: Consensual Approaches to Resolving Public Disputes*. New York: Basic Books.

Susskind, L., S. McKearnen, and J. Thomas-Lamar, eds. 1999. *The Consensus Building Handbook: A Comprehensive Guide to Reaching Agreement*. Thousand Oaks, CA: Sage.

Susskind, L., and L. Zion. 2002. *Can America's Democracy Be Improved?* Cambridge, MA: Consensus Building Institute.

Sutherland, W. J., R. Aveling, T. M. Brooks, M. Clout, L. V. Dicks, L. Fellman, E. Fleishman, et al. 2014. "Horizon Scan of Global Conservation Issues for 2014." *Trends in Ecology and Evolution* 29, no. 1: 15–22.

Sutherland, W. J., and H. J. Woodroof. 2009. "The Need for Environmental Horizon Scanning." *Trends in Ecology and Evolution* 24, no. 10: 523–27.

Taleb, N. N. 2007. *The Black Swan: The Impact of the Highly Improbable Fragility.* New York: Random House.

Tansley, A. G. 1935. "The Use and Abuse of Vegetational Concepts and Terms." *Ecology* 16, no. 3: 284–307. doi:10.2307/1930070.

Tarantola, A. 2006. "Popper, Bayes and the Inverse Problem." *Nature Physics* 2: 492–94.

Teper, I. 2014. "Inconstants of Nature." *Nautilus.* http://nautil.us/issue/9/time /inconstants-of-nature.

Terborgh, J., and J. Estes. 2010. *Trophic Cascades: Predators, Prey and the Changing Dynamics of Nature.* Washington, DC: Island Press.

Tett, P., R. J. Gowen, S. J. Painting, M. Elliott, R. Forster, D. K. Mills, E. Bresnan, et al. 2013. "A Framework for Understanding Marine Ecosystem Health." *Marine Ecology Progress Series* 494: 1–27. doi:10.3354/meps10539.

Thompson, J. N. 1998. "Rapid Evolution as an Ecological Process." *Trends in Ecology and Evolution* 13: 329–32.

Thompson, W. C. 1989. "Are Juries Competent to Evaluate Statistical Evidence?" *Law and Contemporary Problems* 52, no. 4: 9–41.

Tierney, K., and M. Bruneau. 2007. "Conceptualizing and Measuring Resilience: A Key to Disaster Loss Reduction." *TR News* 250: 14–17.

Tilman, D. 1996. "Biodiversity: Population versus Ecosystem Stability." *Ecology* 77: 350–63.

———. 1999. "The Ecological Consequences of Changes in Biodiversity: A Search for General Principles." *Ecology* 80: 1455–74.

Tilman, D., and C. Lehman. 2001. "Human-Caused Environmental Change: Impacts on Plant Diversity and Evolution." *PNAS* 98, no. 10: 5433–40. doi:10.1073/pnas.091093198.

Tilman, D., and S. Pacala. 1993. "The Maintenance of Species Richness in Plant Communities." In *Species Diversity in Ecological Communities: Historical and Geographical Perspectives*, edited by R. E. Ricklefs and D. Schluter, 13–25. Chicago: University of Chicago Press.

Torrens, P. M., and M. Alberti. 2000. *Measuring Sprawl.* CASA Working Papers 27. London: Centre for Advanced Spatial Analysis. www.casa.ucl.ac.uk /measuringsprawl.pdf.

Torres-Sosa, C., S. Huang, and M. Aldana. 2012. "Criticality Is an Emergent Property of Genetic Networks That Exhibit Evolvability." *PLOS Computational Biology* 8. doi:10.1371/journal.pcbi.1002669.

Turner, B. L. II, and W. B. Meyer. 1993. "Environmental Change: The Human

Factor." In *Humans as Components of Ecosystems: The Ecology of Subtle Human Effects and Populated Areas*, edited by M. J. McDonnell and S. T. A. Pickett, 40–50. New York: Springer.

Turner, M. G. 1987. *Landscape Heterogeneity and Disturbance. Ecological Studies* 64. New York: Springer.

Turner, M. G., S. R. Carpenter, and E. J. Gustafson. 1998. "Land Use." In *Status and Trends of the Nation's Biological Resources*, edited by M. J. Mac, P. A. Opler, and C. E. P. Haecker, 1: 27–61. Washington, DC: U.S. Department of the Interior, Geological Survey.

Turner, M. G., R. H. Gardner, and R. V. O'Neill. 2001. *Landscape Ecology in Theory and Practice: Pattern and Process*. New York: Springer.

Tzoulas, K., K. Korpela, S. Venn, V. Yli-Pelkonen, A. Kaźmierczak, J. Niemela, and P. James. 2007. "Promoting Ecosystem and Human Health in Urban Areas Using Green Infrastructure: A Literature Review." *Landscape and Urban Planning* 81, no. 3: 167–78.

Ulanowicz, R. E. 1986. *Growth and Development: Ecosystems Phenomenology*. San Jose, CA: Excel Press.

UN-Habitat. 2008. *State of the World's Cities 2008/2009: Harmonious Cities*. London: Earthscan Publications.

United Nations. 2011. *World Urbanization Prospects: The 2011 Revision, Highlights (ST/ESA/SER.A/322)*. Department of Economic and Social Affairs, Population Division. http://www.un.org/en/development/desa/population/publications/pdf/urbanization/WUP2011_Report.pdf.

———. 2013. *State of the World's Cities 2012/2013: Prosperity of Cities*. New York: Routledge. https://sustainabledevelopment.un.org/content/documents/745habitat.pdf.

———. 2014. *World Urbanization Prospects: The 2014 Revision, Highlights (ST/ESA/SER.A/352)*. New York: United Nations, Department of Economic and Social Affairs, Population Division.

United Nations Framework Convention on Climate Change (UNFCCC). 2015. Conference of the Parties (COP) 2015. "Adoption of the Paris Agreement. Proposal by the President." Paris, December 12, 2015. Geneva: United Nations.

Van den Hove, S. 2007. "A Rationale for Science-Policy Interfaces." *Futures* 39, no. 7: 807–26.

Van der Leeuw, S. E. 2007. "Information Processing and Its Role in the Rise of the European World System." In *Sustainability or Collapse? An Integrated History and Future of People on Earth*, edited by R. Costanza, L. J. Graumlich, and W. L. Steffen, 213–41. Cambridge, MA: MIT Press.

Van der Leeuw, S., R. Costanza, S. Aulenbach, S. Brewer, M. Burek, S. Cornell, C. Crumley, et al. 2011. "Toward an Integrated History to Guide the Future." *Ecology and Society* 16, no. 4: 2. doi:10.5751/ES-04341-160402.

Van Kerkhoff, L., and L. Lebel. 2006. "Linking Knowledge and Action for Sustainable Development." *Annual Review of Environment and Resources* 31: 445–77.

van Vliet, M., K. Kok, A. Veldkamp, and S. Sarkki. 2012. "Structure in Creativity: An Exploratory Study to Analyse the Effects of Structuring Tools on Scenario Workshop Results." *Futures* 44, no. 8: 746–60. http://dx.doi.org/10.1016/j.futures.2012.05.002.

Vicino, T. J., B. Hanlon, and J. R. Short. 2007. "Megalopolis 50 Years On: The Transformation of a City Region." *International Journal of Urban and Regional Research* 31, no. 2: 344–67. doi:10.1111/j.1468-2427.2007.00728.x.

Vietz, G. J., C. J. Walsh, and T. D. Fletcher. 2015. "Urban Hydrogeomorphology and the Urban Stream Syndrome: Treating the Symptoms and Causes of Geomorphic Change." *Progress in Physical Geography.* doi:10.1177/0309133315605048.

Vitousek, P. M., L. A. Mooney, J. Lubchenco, and J. M. Melillo. 1997. "Human Domination of Earth's Ecosystems." *Science* 277: 494–99.

von der Lippe, M., J. M. Bullock, I. Kowarik, T. Knopp, and M. Wichmann. 2013. "Human-Mediated Dispersal of Seeds by the Airflow of Vehicles." *PLOS ONE* 8, no. 1: e52733. doi:10.1371/journal.pone.0052733.

Wack, P. 1985. "Scenarios: Uncharted Waters Ahead." *Harvard Business Review* 63: 72–89.

Wagner, A. 2005. *Robustness and Evolvability in Living Systems.* Princeton, NJ: Princeton University Press.

Waide, R. B., M. R. Willig, C. F. Steiner, G. Mittelbach, L. Gough, S. I. Dodson, G. P. Juday, and R. Parmenter. 1999. "The Relationship between Productivity and Species Richness." *Annual Review of Ecology and Systematics* 30: 257–301.

Walker, B., C. S. Holling, S. R. Carpenter, and A. Kinzig. 2004. "Resilience, Adaptability and Transformability in Social-Ecological Systems." *Ecology and Society* 9, no. 2: 5.

Walker, B., and J. A. Meyers. 2004. "Thresholds in Ecological and Social-Ecological Systems: A Developing Database." *Ecology and Society* 9, no. 2: 3.

Walker, W. E., M. Haasnoot, and J. H. Kwakkel. 2013. "Adapt or Perish: A Review of Planning Approaches for Adaptation under Deep Uncertainty." *Sustainability* 5: 955–79.

Walsh, C. J., T. D. Fletcher, and A. R. Ladson. 2005. "Stream Restoration in Urban Catchments through Redesigning Stormwater Systems: Looking to the Catchment to Save the Stream." *Journal of the North American Benthological Society* 24: 690–705.

Walsh, M. R., J. P. DeLong, T. C. Hanley, and D. M. Post. 2012. "A Cascade of Evolutionary Change Alters Consumer-Resource Dynamics and Ecosystem Function." *Proceedings of the Royal Society of London B* 279: 3184–92.

Walters, A. W., R. T. Barnes, and D. M. Post. 2009. "Anadromous Alewives, no. *Alosa Pseudoharengus*) Contribute Marine-Derived Nutrients to Coastal Stream Food Webs." *Canadian Journal of Fisheries and Aquatic Sciences* 66, no. 3: 439–48.

Wandeler, P., S. M. Funk, C. R. Largiader, S. Gloor, and U. Breitenmoser. 2003.

"The City-Fox Phenomenon: Genetic Consequences of a Recent Colonization of Urban Habitat." *Molecular Ecology* 12: 647–56.

Warwick, R. M., S. L. Dashfield, and P. J. Somerfield. 2006. "The Integral
Structure of a Benthic Infaunal Assemblage." *Journal of Experimental
Marine Biology and Ecology* 330: 12–18.

Webster, J. R., and E. F. Benfield. 1986. "Vascular Plant Breakdown in Freshwater Ecosystems." *Annual Review of Ecology and Systematics* 17, no. 1:
567–94.

West, G. B., and J. H. Brown. 2005. "The Origin of Allometric Scaling Laws in
Biology from Genomes to Ecosystems: Towards a Quantitative Unifying
Theory of Biological Structure and Organization." *Journal of Experimental
Biology* 208: 1575–92.

West, G. B., J. H. Brown, and B. J. Enquist. 1997. "A General Model for the Origin of Allometric Scaling Laws in Biology." *Science* 276, no. 5309: 122–26.
doi:10.1126/science.276.5309.122.

———. 1999. "The Fourth Dimension of Life: Fractal Geometry and Allometric Scaling of Organisms." *Science* 284: 1677–79.

Westley, F., P. Olsson, C. Folke, T. Homer-Dixon, H. Vredenburg, D. Loorbach,
J. Thompson, et al. 2011. "Tipping Toward Sustainability: Emerging Pathways of Transformation." *Ambio: A Journal of the Human Environment* 40,
no. 7: 762–80. doi:10.1007/s13280-011-0186-9.

Weyl, H. 1949. *Philosophy of Mathematics and Natural Science.* Princeton, NJ:
Princeton University Press.

White, P. S., and S. T. A. Pickett. 1985. "Natural Disturbance and Patch
Dynamics: An Introduction." In *The Ecology of Natural Disturbance and
Patch Dynamics,* edited by S. T. A. Pickett and P. S. White, 3–13. Orlando,
FL: Academic Press.

Whitehead, A., A. Triant, D. Champlin, and D. Nacci. 2010. "Comparative
Transcriptomics Implicates Mechanisms of Evolved Pollution Tolerance
in a Killifish Population." *Molecular Ecology* 19, no. 23: 5186–203.

Whitham, T. G., J. K. Bailey, J. A. Schweitzer, S. M. Shuster, R. K. Bangert,
C. J. LeRoy, E. V. Lonsdorf, et al. 2006. "A Framework for Community and
Ecosystem Genetics: From Genes to Ecosystems." *Nature Reviews Genetics*
7: 510–23.

Wiesmann, U., S. Biber-Klemm, W. Grossenbacher-Mansuy, and G. Hirsch
Hadorn. 2008. "Enhancing Transdisciplinary Research: A Synthesis in
Fifteen Propositions." In *Handbook of Transdisciplinary Research,* edited by
G. Hirsch Hadorn, H. Hoffmann-Riem, S. Biber-Klemm, W. Grossenbacher-
Mansuy, D. Joye, C. Pohl, U. Wiesmann, and E. Zemp, 433–41. Dordrecht:
Springer.

Wilby, R. L., S. P. Charles, E. Zorita, B. Timbal, P. Whetton, and L. O. Mearns.
2004. *Guidelines for Use of Climate Scenarios Developed from Statistical Down-
scaling Methods.* IPCC Task Group on Data and Scenario Support for Impact
and Climate Analysis. Norwich, UK: TGCIA.

Williams, J. G., R. W. Zabel, R. S. Waples, J. A. Hutchings, and W. P. Connor. 2008. "Potential for Anthropogenic Disturbances to Influence Evolutionary Change in the Life History of a Threatened Salmonid." *Evolutionary Applications* 1, no. 2: 271–85. doi:10.1111/j.1752-4571.2008.00027.x

Wise, R. M., I. Fazey, M. Stafford Smith, S. E. Park, H. C. Eakin, E. R. M. Archer Van Garderen, and B. Campbell. 2014. "Reconceptualising Adaptation to Climate Change as Part of Pathways of Change and Response." *Global Environmental Change* 28: 325–36.

Wolman, M. G. 1967. "A Cycle of Sedimentation and Erosion in Urban River Channels." *Geografiska Annaler,* series A, *Physical Geography* 49: 385–95.

World Economic Forum. 2015. *Top 10 Urban Innovations.* https://agenda .weforum.org/2015/10/top-10-urban-innovations-of-2015/. Accessed October 2015.

Wright, F. L. 1932. *The Disappearing City.* New York: W. F. Payson.

Wu, J., and O. L. Loucks. 1995. "From Balance of Nature to Hierarchical Patch Dynamics: A Paradigm Shift in Ecology." *Quarterly Review of Biology* 70: 439–66.

Yeh, P. J., and T. D. Price. 2004. "Adaptive Phenotypic Plasticity and the Successful Colonization of a Novel Environment." *American Naturalist* 164: 531–42.

Yohe, G., and R. Leichenko. 2010. "Adopting a Risk-Based Approach." *Annals of the New York Academy of Sciences* 1196: ch. 2, 29–40.

Young, O. R., F. Berkhout, G. C. Gallopin, M. A. Janssen, E. Ostrom, and S. van der Leeuw. 2006. "The Globalization of Socio-Ecological Systems: An Agenda for Scientific Research." *Global Environmental Change* 16, no. 3: 304–16.

Index

cities that think like planets, 207–24; diversity of solutions, 214; imagination of future cities, 3–10, 214–17; principles of, 214

city planning. *See* urban planning

climate agreement (Dec. 2015), 213–14, 219

climate change: adaptation, elements of, 221–22; adaptation planning, 103; boom-and-bust vs. stable economy, 127–30, 128–30*figs.*; Copenhagen 2105 Climate Plan, 219; extreme climate events, 85–86, 89*box*, 93*fig.*, 96, 131; flexible adaptation pathways, 222–23*box*; flooding events and, 88*box*; human responses to, 43; human role in, 213; hypothetical scenarios, 127–30*figs.*; IPCC report on, 96; planetary boundaries and, 210; planning and adapting for, 219–21; rapid or abrupt, 209; regime shifts due to, 64; risk management and, 96; risks of, differences across regions, 217–18; urbanization and, 91–92. *See also* regime shifts; tipping points

coastal regions: extreme events and, 93, 96; regime shifts in, 87–88*box*, 87*fig.*, 93, 95–96; sea-level rise and, 96

co-evolution, xii, 48–51, 142, 208; adaptive, 40; adaptive capacity and, 48–51, 49*table*, 52*fig.*; complex systems and, 31; feedbacks and, 50–51, 159; of human and natural systems, 11, 24, 27*fig.*, 28–29, 31, 40, 148, 158–59, 212–13; of hybrid ecosystems, 158–59; networks in, 22–23; of novel ecosystems, 25–26, 26*fig.*, 27*fig.*; in planning paradigms, 49, 49*table*; resilience and, 48–51, 49*table*, 82–83, 106, 212; in urban ecosystems, 49*table*, 50–51,

50*fig.*, 78. *See also* eco-evolution; evolution

Collins, J., 63

communication network, 86

complexity, 24, 30–33, 132, 162; adaptive capacity and, 108, 215–17, 216*fig.*; of cities, 12, 30–33; in hybrid ecosystems, 30–33, 32*fig.*, 52, 52*fig.*, 87; study of, 101; tipping cascades and regime shifts and, 209; unpredictability and, 29; in urban ecosystems, 52, 52*fig.*, 64, 65–68, 66–67*figs.*, 87. *See also* networks

complex societies, evolution of, 213

complex systems: adaptive capacity and innovation in, 90, 215–17, 216*fig.*; paradigm, 64; resilience properties in, 40–41, 215–17, 216*fig.*; theory, 98

connectivity, 215, 219

cooperation, 15; evolution of, 213; social networks and, 17

Copenhagen, Denmark, 172, 219

coupled human and natural systems, 19–20, 79–80; adaptive capacity of, 29, 48–51, 51*fig.*; cities as, 11, 42–43; dynamics of, 25–26, 27*fig.*, 28–29; emergent properties in, 84–105; hybrid nature and processes, 213; hybrid networks and, 21; inverse experiments and, 164–67; planning paradigms, 49*table*; properties enhancing adaptive capacity and innovation, 41; resilience of, 82–83; scientific challenges of, 51–57, 52*fig.*; transformation and, 49, 49*table*, 55; uncertainty in, 97–98. *See also* co-evolution

Crepis sancta, 141*table*, 145*box*, 145*fig.*

criticality, 86; self-organized, 90

critical state/point, 41–42, 90

critical systems, 41–42, 86

critical transitions, 39, 47, 108, 162; anticipation of, 89*box*, 171
cross-scale interactions, 21–24, 51*fig.*, 190–91, 215–16, 216*fig.*; adaptive capacity and, 50–51, 52*fig.*, 90, 215; amplifications of, 47*box*; functional redundancy from, 91; in hybrid ecosystems, 21–24, 38, 39; innovation and, 22, 41, 51*fig.*; lag time in, 42; opportunities from, 47*box*; in planetary boundary approach, 210; resilience and, 41, 213, 215–16, 216*fig.*
Crutzen, P. J., x, 11, 91, 92*fig.*, 213, 225
Cumming, G. S., 203–4

D

Da Nang, Vietnam, 173
Daphnia, 145*fig.*; trait changes and urban signature, 140*table*, 144, 144*box*
dark-eyed junco, 140*table*, 144*box*, 145*fig.*
data-mining technologies, 174
decentralization, 39
decision making: anticipation of critical events and effects, 89*box*; context of, 205; experience of time and space, influence on, 208; incomplete knowledge and, 180; stakeholder access to, 105; time scales and, 102–3, 214–15; uncertainty and, 102, 172, 180, 223, 224
deductive reasoning, 167
delta cities, 31, 34
deltas, river, 87*box*, 88*box*, 96
differentiation, adaptation/innovation and, 38–39, 213
discontinuity hypothesis, 81–82
disturbance, 64, 73–74; ecosystem functions and, 59; Intermediate Disturbance Hypothesis, 73; novel disturbances, 147, 152–53

diversity, 130–31, 131*fig.*, 214, 215, 216*fig.*, 218; adaptive capacity and, 215; genetic diversity, cities and, 77; species diversity, 69–71. *See also* biodiversity

E

early warning systems, 216, 216*fig.*, 221; adaptive capacity and innovation and, 41, 90; importance of, 89*box*
Earth: abrupt changes and regime shifts, 209; age of, 208; evolution of, role of cities in, 207–8. *See also* tipping points
earthworms, 133; trait changes and urban signature, 140*table*, 145*fig.*, 146*box*
eco-evolution, 133–63; change, scenarios of, 160–61, 161*fig.*; dynamics of, 61–62, 138*fig.*; examples of, 139–41*table*, 143–44, 144–46*box*; genetic signatures in, 135, 142–44; human roles and impacts, 133–35, 137–42, 139–41*table*; hybrid nature of ecosystems, 142; innovation in, 142, 159–62; Palkovacs and Hendry (2010) framework, 137, 138*fig.*; phenotypic change, 154–58; phenotypic traits, 135, 139–41*table*; rapid change, 133; trait changes, 135, 139–41*table*; urban eco-evolution, 142–44, 143*fig.*, 144–46*box*. *See also* co-evolution; evolution
eco-evolutionary dynamics, 61–62, 137–42, 138*fig.*; co-evolution and, 142, 148; human roles and feedbacks, 142; humans, integrating into, 137–42; hypotheses on, 138–42
eco-evolutionary feedbacks, 61–62, 77–78, 134–37, 155–57*table*, 159;

ecosystem functions, effects on, 135; examples of, 139–41*table*; human role in, 134, 135, 142, 147–53

eco-evolutionary mechanisms, 146–58; biotic interactions, 151–52; habitat modification, 150*fig.*, 151; heterogeneity, 152; homogenization hypothesis, 147–48*box*; mechanisms by which evolutionary change affects urban ecosystems, 154–58; mechanisms by which urbanization affects evolutionary dynamics, 147–53; novel disturbance, 152–53; social interactions, 153; tele-coupling, 153; urban gradient, 149–50*box*, 150*fig.*

eco-evolutionary networks, 21

ecological memory, 39

ecological niche. *See* niche

ecological resilience, 39–40, 40*fig.*, 109. *See also* resilience

ecology, 58–61; in the Anthropocene, 60; including humans in consideration of, x, 25–26, 48*fig.*, 53–54, 58, 60–61, 69; new paradigm for, 60–61

economic conditions: climate change scenarios under, 127–30, 128–30*figs.*; microcredit, 173

economies of scale, 13, 19

ecosystem functions, 30, 33*fig.*, 38, 65*fig.*, 68, 78–79; disturbance regimes and, 59; eco-evolutionary feedbacks and, 135; heritable change associated with, 155–57*table*; heritable traits and, 154–58, 155–57*table*, 158; humans and, 59, 61, 64, 68, 69, 78–79, 124–25, 125*fig.*; human well-being as, 65*fig.*, 135; at multiple scales of time and space, 51; species diversity and, 69–70; urban development patterns and, 114–25, 135

ecosystem management, 25

ecosystem resilience: conceptual framework, 62–68, 65*fig.*; humans and, 64

ecosystems: discontinuities in, 38, 81–82; dynamics of, 32, 33*fig.*, 80; ecosystem concept, 59; historical state, 25, 26*fig.*, 27*fig.*, 39; humans as external to, 25, 43, 60–61, 68; humans included in, x, 25–26, 48*fig.*, 53–54, 58, 60–61, 63, 78–79, 104; hybrid, 11–57, 26*fig.*, 27*fig.*; incomplete knowledge and, 80; multiple equilibria, 79, 80; novel, 24–26, 26*fig.*, 27*fig.*, 39–40; paradigm shift for, 25, 78–83; self-regulation of, 59. *See also* hybrid ecosystems

ecosystem services, 78–79, 135

Einstein, Albert, 3

electric grid, 23; modularity in, 91, 215

emergence of, cities, 43

emergent properties, 34–35, 84–105; adaptive capacity and, 50, 50*fig.*; of cities, 38; framework and questions for urban ecology, 98–99; of hybrid ecosystems, 32*fig.*, 34–35, 34*fig.*, 64; infrastructure and urban resilience, 87–90*box*, 87–88*figs.*; innovation, 96–97; knowledge synthesis, 103–5; planetary boundaries and, 210; regime shifts in urban ecosystems, 91–96; research agenda and questions, 99–102; resilience, 90–98, 106; self-organized criticality, 90; of social systems, 19–20, 23, 64; uncertainty, 97–98; of urban ecosystems, 38, 84–90

Enquist, B. J., 15, 16, 82

environmental change, cities as drivers of, 46*box*, 47*box*

equilibria, multiple, 60

estuaries, 31–33, 37*fig.*
ethics in hybrid planet, 224
eutrophication, 97, 113, 113*fig.*
evolution: drivers of, 212; of Earth, role of cities in, 207–8; ecological change scenarios, 160–61, 161*fig.*; fitness landscape and, 40; Great Oxidation Event, 60, 162, 207; humans as drivers of, xiv, 25, 78, 133, 135; microevolutionary change, human role in, 77, 78, 133–35, 151; natural selection and, 17–18, 212–13; of networks, 17–19; phenotypic, 154–58, 155–57*table*; rapid, 61–62, 133, 135, 136; role of humans in, 25, 40, 62, 77, 208, 213; scenarios of evolutionary change, 161*fig.*; and short-term ecological change, 61–62; urban evolutionary change, detecting, 142–44, 143*fig.*; urbanization, effects on evolutionary dynamics, 147–53. *See also* co-evolution; eco-evolution
exotic species, 151
experiments: designing, 167–70; learning through, 218; real-world problems, 175–77; reverse experiments, 164–78, 218. *See also* reverse experiments
external shocks, 29, 36–37*figs.*, 37
extinctions, mass, 213
extreme events, 85–86, 89*box*, 93*fig.*, 96, 131

F

fast variables, 87*fig.*, 88*box*, 113*fig.*, 178*box*
Fedora, xii, 9, 10
feedbacks, 25–26, 50–51, 99; changes in, 42, 43, 60; early warning and, 216, 216*fig.*; eco-evolutionary, 61–62, 77–78, 134–37; evolution of systems and, 43; feedback loops, 86;

increase in Anthropocene, 60, 207; novel/emergent, 34–35, 35*fig.*; regime shifts and, 36, 36*fig.*, 87*box*, 94, 95; research on, 99–100; in urban ecosystems, 85–86
Feldman, M. W., 62, 134, 154
financial instruments, 221
financial networks, 23
fish, 133; trait changes and urban signature, 139*table*, 143, 144*box*
fitness landscape, 40
flexibility, heterogeneity and, 90–91, 101
flexible adaptation pathways, 222–23*box*
flood control strategies, 33, 35*fig.*; emergent feedback and, 35, 35*fig.*; infrastructure, 85–86, 95*fig.*; London Thames barriers, 222*box*; the Netherlands, 220; New York City, 222–23*box*; Venice, Italy, 54–57*box*; and vulnerability to extreme climate events, 85–86, 96
flood discharge, impervious surfaces and, 74–75, 115
Folke, C., 229
Forman, R. T. T., x, xii, 61
4 Rs, 89–90*box*
fragmentation, 142

G

Geddes, P., 12
gene frequencies, 135
genetic diversity, 77
genetic signatures, 77, 135, 142
Gleason, H. A., 24
global economy, urban agglomerations and, 45*box*
globalization, time-space compression and, 45*box*
Gottmann, Jean, 45*box*
government, 80; emerging mega

regions and, 47*box*; implementation of plans by, 103; top-down decisions, 23, 68, 105. *See also* local governments
Great Oxidation Event (GOE), 60, 162, 207
green infrastructure, 21, 211
green space, 21
Grimm, N. B., x, xii, 61, 63, 75, 107
Groffman, P. M., 68–69, 71, 74, 119, 147, 152
Gunderson, L., xi, 38, 64, 82, 84, 190, 213, 226

H

habitats: alteration of, 142, 146, 147, 147–48*box*, 150*fig.*, 155–57*table*; fragmentation of, 142; genetic diversity, effect on, 77; heterogeneity, 147, 147–48*box*, 152; human, 11, 20, 25, 27*fig.*, 48*fig.*, 68–69; loss of, 46*box*, 151; novel, 159; urban, 68–69, 72, 77–78, 142
Harris, J. A., 25, 26*fig.*, 227
heat islands, 107, 157*table*
Hendry, A. P., 77, 135, 137, 138, 138*fig.*, 154, 226
heritable traits, 155–57*table*
heterogeneity, 101, 152, 215, 216*fig.*, 218; adaptive capacity and, 50, 50*fig.*, 90–91, 108, 215; critical transitions and thresholds and, 108; flexibility and, 90–91; in hybrid cities, 219; in hybrid networks, 22, 23; scale and, 169–70; spatial, in urban regions, 70, 147–48*box*, 169
hierarchical, branching networks, 16, 101, 213
Higgs, E., 25, 26*fig.*, 227
highly optimized tolerance (HOT) theory, 126
Hirsch, Paul, xii, 224
Hobbs, R. J., 24–26, 26*fig.*, 227

Holling, C. S., xi, 29–30, 38, 39, 42, 59, 64, 80–82, 84, 98, 108–9, 158, 184–85, 196, 213, 225, 226, 228, 229
homogenization hypothesis, 147–48*box*
Huang, S., 42, 160
human action/agency, 23, 24–25, 32*fig.*, 104
human-driven trait changes, 135, 139–41*table*, 143, 144–46*box*
human function, in human vs. natural habitats, 48*fig.*
human habitats, 11, 20, 25, 27*fig.*, 48*fig.*, 68–69
human networks, 17–19, 18*fig.*
humans: as drivers of change, xiv, 25, 54–57*box*, 65*fig.*, 69, 78, 91, 92*fig.*, 104, 124, 135; effects on species diversity and selection, 69–70, 71–72; as external agents in ecosystems, 25, 43, 60–61, 68; inclusion in consideration of ecology/ecosystems, x, 25–26, 48*fig.*, 53–54, 58, 60–61, 63, 69, 78–79, 104, 136, 137–42, 163; interactions between human and evolutionary processes, xiv, 139–41*table*, 147–53; novel conditions arising from, 25, 159; as part of urban ecology, x, 63, 69; perceptions and behaviors and, 124; role in climate change, 213; role in eco-evolutionary dynamics, 61–62, 134–42; role in evolution, 40, 62, 208, 213. *See also* Anthropocene; coupled human and natural systems
human societies, evolution of, 19–20
human systems, co-evolution with natural systems, 27*fig.*, 28–29, 40, 212–13
human well-being: defining and assessing, 52; ecosystem and, 65*fig.*; network structure and, 66–67*figs.*

tive capacity and, 90; characteristics of innovating systems, 41,
213, 215; cities and, xi, 27; coupled
human and natural systems and,
41; cross-scale interactions and, 22,
41, 51*fig.*; eco-evolution and, 142,
159–62; as emergent property,
96–97; framework for study, 99;
hybrid ecosystems and, 30, 32*fig.*,
38–39, 142, 162, 207–8; institutional, 103; interactions leading
to, 96–97; planning principles to
enable, 218; resilience and, 90–
98, 213, 218; social interactions
and, 20, 27; socioecological, 30,
96–97; sustainable futures and,
211–12

instability: of hybrid systems, 213;
resilience and, 213

institutions: change in, 171; innovation, 103; novel frameworks for,
221; sustainable, 103

interactions. *See* cross-scale interactions; novel interactions

interconnectedness, 47*box*, 86, 215,
219

interdependence, 86, 185, 221; tipping cascades and regime shifts
and, 86, 209; vulnerability and,
86

intertemporality, 221

inverse modeling, 177–78*box*

inverse problems, 164–67; causality
in, 166–67; history in, 166; measurements in, 166

Invisible Cities (Calvino), xii, 10

irreversibility, threshold of, 25

Ives, A. R., 168

local governments, 219; financing mechanisms, 221; infrastructure decisions, 217

London, UK, 172; Thames barriers, 222*box*

long now, 207

Long Term Ecological Research (LTER), 69–70

Lynch, Kevin, 12, 208

M

Marsh, G. P., 59

Marzluff, John, 109, 121, 123, 227

Matthews, B., 135–36, 144, 154

Mayan cities, collapse of, 106

McDannald, M. A., 20–21

McDonnell, M. J., x, xii, 60, 64, 85, 229

McHale, M. R., 43, 45, 47

McKinney, M. L., 68, 70, 147, 152

McNeill, J. R., 91, 92*fig.*

McPhearson, T., 60

measurements, 52*fig.*, 53–55

megacities, 47*box*

megalopolis, 45*box*, 46*fig.*

megaregions, 45–46*box*, 46*fig.*, 47*box*

memory, ecological, 39

metropolis, 46*fig.*, 47*box*

mice, 141*table*, 145*box*, 145*fig.*

microcredit, 173

minerals, 75–77

Mississippi River, channel migration in, 34, 34*fig.*

models, 196, 224. *See also* scenarios

modular electric grid, 215

modularity, 101, 215, 216*fig.*; adaptation/innovation and, 38–39, 213; adaptive capacity and, 50, 50*fig.*, 90, 91, 108, 215; autonomous function and, 91, 101; critical transitions and thresholds and, 108; in hybrid cities, 219; in hybrid

networks, 22, 23; resilience and, 40–41, 91

monitoring, 174–75, 221

Monte Carlo-Markov Chain (MCMC) methods, 167

MOSE Project, 54–57*box*, 56*fig.*

multidimensionality, 221

Multidisciplinary Center for Earthquake Engineering Research (MCEER), 89*box*

multiple equilibria systems, 60, 61, 80

municipal bonds, 221

myths, 185–88

N

natural selection, 17–18, 212–13; humans as selective agents, 135–36, 137; networks governed by, 20–21; new selection pressures, 142

Netherlands: flood control strategies, 220; resilience strategies, 220, 221

net primary productivity, 151

networks, 15–21, 18*fig.*; co-evolutionary, 22–23; complex, 15, 22, 65–68, 66–67*figs.*, 212–13; complexity and interdependence of, 209; differentiation in, 20–21; emergent, 18*fig.*; functions of, 16; hierarchical, branching, 16, 213; hybrid, 16–23, 18*fig.*, 101; natural selection and, 20–21; nodes of, 16, 20–21, 22, 23; novel, 21; robustness in, 22, 23; scaling and, 16; size of, 20–21; social vs. ecological, 17–19, 18*fig.*; structure and dynamics, 65–66, 66–67*figs.*; types of, 18*fig.*; vulnerability to attacks, 23. *See also* social networks; socioeconomic networks

network structure, resilience and, 22

New Orleans: Hurricane Katrina,

Polo, Marco, xii, 9, 10
population size, 12, 38, 43*box*, 45–46, 46*fig.*
Post, D. M., x, 62, 134, 137, 138, 139*table*, 143, 144, 155, 156
power laws, 13–15, 38; critical systems and, 41–42; networks and, 16
predictive models, 53, 128, 129*fig.*
Prigogine, I., 64, 84, 86, 229
probability distributions, 128–29, 129*fig.*
productivity, 151

Q

QECBs, 221

R

"realized niche," Hutchinson's, 62, 72, 163, 227
real-world problems, 175–77
redundancy, 89*box*, 91, 101, 215–16, 216*fig.*
reference conditions, 55; defining, 54–57*box*
regime, definition of, 36, 94
regime shifts, 36–38, 36–37*figs.*, 91–96; abrupt, 36, 92, 97, 209; anticipation of, 89*box*; avoidance of, 210; ball-and-cup diagram of, 36–37*figs.*, 94*fig.*; climate change and, 64; definition of, 93–94, 94*fig.*; drivers of, 93; examples of, 94–95, 97; external shocks triggering, 94, 94*fig.*; extreme events, 93*fig.*; in hybrid ecosystems, 32*fig.*, 36–38, 36–37*figs.*; in hydrological systems, 87–88*box*, 87–88*figs.*, 95*fig.*; irreversible, 92; multiple, 126; prediction of, 184; resilience and, 64, 92–93, 95*fig.*, 100, 103, 126; slow and fast variables, 87–88*box*, 87*fig.*; thresholds for predicting, 48*fig.*; uncertainty

and, 183–84; in urban ecosystems, 91–96
research agenda, 99–102; adaptive capacity, 99; comparative stuey of cities, 99–100, 101; key questions of, 100; resilience, 99–100; socio-ecological interactions and feedbacks, 99–100
resilience, 90–98, 106–32; anticipation of critical events and effects, 89*box*; assessment of, 82, 89–90*box*; of cities, on planetary time scale, 207, 208–9, 214–15; co-evolution of human and natural systems, 42–43, 48–51, 49*table*, 82–83, 212–13; conceptual framework, 62–68, 65*fig.*; decreased, regime shifts and, 64, 100; definition of, 39, 108–9; ecological, 39–40, 40*fig.*, 109; of ecosystems, 82–83; of ecosystems, humans and, 64; as emergent property, 90–98, 105; 4 Rs (robustness, redundancy, resourcefulness, rapidity), 89–90*box*; historical relationships and, 106–7; human perceptions and behaviors and, 124–25, 125*fig.*; of hybrid ecosystems, xi, 23–24, 39–47, 106–32, 213, 219; in hybrid systems, 108–11, 213, 215–17, 216*fig.*; hypothesis for urban patterns, 111–14, 112*fig.*; imagination of a resilient urban planet, 212–14; indicators of, 104; of infrastructure, 42, 89–90*box*; innovation and, 38–39, 213; inverse modeling and, 177–78*box*; New York City example, 222–23*box*; novel resilient systems, 50; predictability, uncertainty, and surprise, 126–30, 132; principles of, 132, 215–17, 216*fig.*; public infrastructure and, 42; reference conditions for, 55; regime shifts and,

48–51, 48*fig.*, 100, 103, 126; reverse experimentation approach, 170; robustness and, 89*box*; scale and, 53, 82; scenarios for, 204–5; of some undesirable states, 48; transferability of, 187; translating concepts into practice, 217–22; urban development patterns and, 111–25, 112*fig.*, 113*fig.*, 130–31, 131*fig.*; varying scales and, 217–18

resilience paradigm, 49, 49*table*

resilience patterns, 125–31; slow and fast variables in, 125–26, 125*fig.*

resilience planning, 48–51, 48*fig.*, 49*table*, 82–83, 131–32, 218, 222–23*box*

resilience principles, 40–42, 132, 215–17, 216*fig.*, 218; cross-scale interactions, 90, 215–16, 216*fig.*; diversity/heterogeneity, 39–40, 90–91, 101, 108, 130–31, 131*fig.*, 215, 216*fig.*; early warning systems, 41, 90, 216, 216*fig.*; modularity, 39–40, 90, 91, 101, 108, 215, 216*fig.*; self-organization, 41, 90, 216*fig.*, 217, 218

resilience theory, 103–4

resilient institutions, 105

response diversity (of species), 136

reverse experiments, 164–78, 218; inverse modeling, 177–78*box*; inverse problems, 164–67; joint fact-finding (JFF), 176; knowledge production, 177; success stories, 170–73

risks: acceptable level of, 222*box*; of climate change, differences cross regions, 217–18; differing perceptions of, 103; uncertainty about, 105

river systems, regime changes in, 109–10, 111*fig.*

robustness, 22, 23, 89*box*, 101, 150, 170

Rome, ancient, 107

Ross, C., 43, 45, 47

Rotterdam, 172

S

salmon: health, network structure and, 67*figs.*; trait changes and urban signature, 139*table*

Sand Engine, The, 220

sand pile, behavior of, 41

scales: in cities that think like planets, 214–15; cross-scale interactions, 21–24, 38; discontinuities in scale, 38, 81; economies of scale, 13, 19; mismatches of scale, 52–53, 52*fig. See also* cross-scale interactions; planetary scales

scaling relationships, 12–15, 82; in cities, 12–15, 15*fig.*, 153; resilience and, 82; scale-free phenomena, 14, 15; superlinear vs. sublinear, 14, 19, 20, 153

scenarios, 192–206, 223–24; alternative, 178*box*; benefits of use, 224; challenges of, 195–96; challenging assumptions, 205–6; climate change, 127–30*figs.*; decision context, 205; ecological change, 160–61, 160*fig.*; of evolutionary change, 161*fig.*; focus on resilience, 204–5; objective of planning with, 223–24; planning elements, actions, and questions, 200–203*table*; risks and opportunities, 206; robust decisions, 206; scenario planning, 104, 200–203*table*, 203–4; steps in and implementation, 198–203; strategic planning and foresight, 196–99, 197*fig.*, 199*fig.*; to study emergent properties, 102; surprise and, 192–93; from two uncertain variables, 127*fig.*; warning signals, 206

Scheffer, M., xi, 22, 29, 40, 50*fig.*, 90, 97, 108, 160, 171, 184, 196, 215, 228
Schoener, T. W., x, 62, 134, 138, 226
scientific challenges, 51–57, 52*fig.*; complexity, 52, 52*fig.*; reference condition, defining, 54–57*box*; scale, 52–53, 52*fig.*; uncertainty, 52*fig.*, 53
sea-level rise: planning defenses against, 220; urban vulnerability to, 96; Venice, Italy, adaptation responses, 54–57*box*
Seattle, Washington: carbon stocks, 115; future scenarios for, 197*fig.*; green areas, 148; Lake Washington eutrophication, 97
seed dispersal, 29, 133–34, 159
selection. *See* natural selection
self-organization, 41, 91, 218; criticality and, 86, 90; of human societies, 16–17; principles of, 90; resilience and, 41, 90, 216*fig.*, 217, 218; role in evolutionary processes, 18, 212
sensory mechanisms, 174–75
Simpson's paradox, 168
slow changes, 87–88*box*, 87*fig.*
slow variables, 87*fig.*, 88*box*, 113, 113*fig.*, 178*box*; definition of, 113
slums, 44
smart design, 221
smart transportation systems, 219, 220*fig.*
social interactions, 148, 153; increase in, 12, 19, 38, 44; innovation and, 20, 27, 38, 97. *See also* cross-scale interactions
social networks, 12, 17; highly-connected nodes of, 22, 23; scaling relationships in, 20; vs. ecological networks, 17–19, 18*fig.*; vulnerability to attacks, 23
social processes, 221–22
socioecological innovation, 30, 96–97

socioecological interactions, 99–100
socioeconomic networks, complexity and interdependence of, 209
solution-oriented research, 211–12
space: human experience of, 208; scales for, 80, 208, 214, 224; urban planning and, 190–91, 214–15
spatial heterogeneity, 70, 169
speciation, 146. *See also* natural selection
species: diversity, 69–71, 136, 152; diversity, disturbance and, 73; exotic, 151; humans as selective agents, 135–36, 137; native, decline in, 120–21, 122, 149*box*; niche construction and, 72; novel interactions of, 213; species-area relationships, 72–73; trait changes, 135, 139–41*table*, 149*box*
spiders, 145*fig.*, 146*box*
sprawl, 84–85
stability, 185–86; management to achieve, 108
Starfield, A. M., 24
Steffen, W., 91, 92*fig.*, 210
stochastic modeling, 104
Stoermer, E. F., 11, 213
succession, 60
superlinear scaling, 14, 19, 20, 153
surprise, 183–84, 192–93; effects on, 88–89*box*
sustainable futures, 211–12
synthesis, 102, 103–5, 167; new synthesis, 134–37

T

Tang, C., 41, 86
Tansley, A. G., 59
technologies, alternative, sustainable futures and, 211–12
telecommunications, 47*box*
tele-coupling, 22, 47*box*, 153, 162
thinking like a planet, xii, 224; imag-

ing a resilient urban planet, 212–
14; imaging future cities, 3–10,
214–17; planetary challenges and
opportunities, 209–12; planetary
time scales, 207, 208–9; reframing
questions for, 222
threshold effects, 221
thresholds: critical, 104, 108, 171;
instability of, 185–86; system
transition, 113; in urban ecosystems, 39, 41*fig.*, 43, 109, 110*fig.*,
112*fig.*, 113
time: human experience of, 208, 218;
intertemporality, 221; time-space
compression, 45*box*
time lags. *See* lag time
time scales: adaptive cycle and, 51,
51*fig.*; decision making and, 102–
3, 214–15; planetary, 207, 208–9,
214–15, 224; urban planning and,
190–91, 208, 214–15
tipping cascades, 162, 209
tipping points: anticipating and identifying, 89*box*, 104; cascades and,
162, 209; feedback increases and,
60; flood events and, 88*box*, 88*fig.*;
misconceptions about, 210; planetary scale, 207, 210; risk of crossing,
209; urban agglomeration and,
43–47*box*, 46*fig.*
top-down interventions, 23
top-down management, complex
urban systems and, 68, 105
Torres-Sosa, C., 42, 160
trade-offs, 51, 79–80
trait changes, 135, 139–41*table*; urban
patterns and, 149*box*
transformability, challenging
assumptions and, 218
transformation, 132; co-evolution
of human and natural systems
and, 82–83; in coupled human and
natural systems, 49, 49*table*, 55;
cross-scale structures and, 38;

opportunities for, 96; potential
triggers for, 60; reference conditions for, 55
transportation: alternatives during
Hurricane Sandy, 131; carbon
emissions and, 117, 125; smart
transportation systems, 219,
220*fig.*
tree cover, models for, 54
tree planting, in urban planning,
54–55
turning points: Great Oxidation
Event (GOE), 60, 162, 207. *See also*
tipping points

U

uncertainty, 52*fig.*, 53, 85, 97–98, 126–
30, 169, 179–91, 182–83, 212, 218;
adaptive capacity and, 44, 218;
climate change and economic
scenarios, 127–30, 128–30*figs.*;
decision making under, 80, 102,
172, 223, 224; in planning paradigms, 49*table*, 102, 132, 188–91;
probability distributions, 128–29,
129*fig.*; resilience in face of, 44,
221; scenarios from two uncertain
variables, 127*fig.*
United States, megaregions in,
45–46*box*
urban agglomeration, 43–47*box*, 61
urban areas, difference in risk and
adaptive capacity, 217–18
urban design, time and space scales
in, 208
urban development patterns, 109,
111–14; bird diversity, 120–24,
122–23*figs.*; carbon stocks/fluxes,
114–18, 117–18*figs.*; compact development, 112, 112*fig.*; dispersed
development, 112, 112*fig.*; diversity
of, 130–31, 131*fig.*; effects on ecosystem functions, 114–25, 149–

and, 47*box*; disturbance regimes, effects on, 73–74; as driver of global-scale phenomena, 91–92; eco-evolutionary affecting, 154–58; eco-evolutionary feedbacks and, 77–78, 135–36; ecological homogenization and, 70–71; eco-system function and, 98–99; effects on evolutionary dynamics, 147–53; emerging urban agglomeration patterns, 44*fig.*; evolutionary change, detecting, 142–44, 143*fig.*; novel interactions of species and, 213; nutrient cycling and, 75–76; population size and, 43*box*, 45–46, 46*fig.*; rapid, 43–45; thresholds and unstable states, 39, 41*fig.*, 43, 109, 110*fig.*, 112*fig.*; tipping points and, 43–47*box*, 46*fig.*; trait changes and, 139–41*table*, 143, 144–46*box*

urban planning: adaptation pathways, 103–4, 207–24; assumptions in, challenging, 53, 217–18; challenges of, 30–31; change, inclusion in planning, 98; co-evolutionary focus and, 217–22; co-production of knowledge and, 104–5; flexible adaptation pathways, 222–23*box*; infrastructure decisions, 97, 98, 105, 132; myth of optimal patterns for, 186–87; planning paradigms, 49, 49*table*, 132; predictive models, 53; principles to enable resilience and innovation, 132; questions for, 100–102, 189–90; for resilience, 131–32; scenario planning, 223–24; synthesis of new knowledge, 102, 103–5, 134–37, 167; time and space scales in, 190–91, 208, 214–15; translating concepts into practice, 217–22; uncertainty and surprise, integrating, 53, 102, 132, 188–91

urban regions: impervious surfaces in, 74–75, 115; vulnerability to sea-level rise, 96
urban rivers, 109–10, 111*fig.*
urban sensors, 174–75
urban signatures, 139–41*table*
urban stream syndrome, 68, 75

V

values, diversity of, 55, 103
variables: fast, 87*fig.*, 88*box*, 113*fig.*, 178*box*; slow, 87*fig.*, 88*box*, 113, 113*fig.*, 178*box*; unknown, 164; urban development patterns and, 112, 113*fig.*
Venice, Italy, 54–57*box*

W

water: consumption and use of, 115–16, 210; urban stream syndrome, 68, 75
water flow: channelization of, 74–75, 95*fig.*; infrastructure for, 107
water pollution, 97, 115; planetary boundaries and, 210
water resources, urbanization effects on, 74–75, 80
wealth production, 13–14, 15*fig.*, 17
weeds, trait changes and urban signature, 141*table*, 145*box*
West, G., 12–13, 15, 16, 135
What Time Is This Place (Lynch), 208
Widders, D., 20–21
Wiesenfeld, K., 41, 86
wildlife, 133
woodpeckers, 123

Z

Zhou, W., 43, 47
Zipf's law, 90